WE ARE AS FLEXIBLE AS RUBBER!

# We are as Flexible as Rubber!

## *Livelihood Strategies, Diversity and the Local Institutional Setting of Rubber Small Holders in Kerala, South India*

BALZ STRASSER

MANOHAR
2009

First published 2009



ISBN 978-81-7304-803-6

*Published by*
Ajay Kumar Jain for
Manohar Publishers & Distributors
4753/23 Ansari Road, Daryaganj
New Delhi 110002

*Typeset by*
Ajay Arts
Delhi 110006

*Printed at*
Salasar Imaging Systems
Delhi 110035

# Contents

# Boxes

# Tables

# Figures

# Acknowledgements

Many people have been part of the journey towards the completion of this study. I am very grateful to Prof. Dr Ulrike Müller-Böker and Dr Urs Geiser of the Human Geography Department, University of Zurich, for their valuable support and supervision throughout the thesis. Thanks are also due to Prof. Dr Hans-Georg Bohle for reviewing it.

Many people assisted and supported me during my fieldwork in Kerala. First of all, I would like to thank Prof. Dr K.N. Nair and Dr Ravi Raman from the Centre of Development Studies in Thiruvananthapuram, as well as Dr Tharian George and Dr A.K. Krishna Kumar from the Rubber Board, for their local assistance and support. I would also like to thank Biju James, Rajan Joseph and Siby for their translation of the interviews, as well as Bobby and Bobban for their survey work. Our stay in Kerala was unmeasurably enriched by the friendship of Rony and Ruby and their family: thank you! I'm also grateful to the whole INFACT team and to all their friends in the farming community for introducing me to the Kerala farmers' way of life. Thanks also for their support and openness to Soman Nair, Vineetha Menon, Antonyto Paul, Vinod, Ramakumar and all others whom I may have forgotten. Finally, I am grateful to the farmers from Thalanadu Panchayat whom I interviewed for sharing their experiences and sometimes amazing livelihood stories.

This work would have been much harder without the moral support of my peers within the NCCR North-South community. Grateful thanks go specially to Christine Bichsel and Silvia Hostettler for sharing both the bitterest and the sweetest moments of recent years, and for their friendship and trust. Thanks also to Ghana Gurung, Tobias Hagmann and to all my other colleagues from the Swiss NCCR North-South PhD network for their open exchange and critique. I would also like to thank all the colleagues of the Human Geography Department for their valuable support in all kinds of administrative, logistical or other matters. Special thanks go to Barbara Zollinger, Jamil Mokthar and Orvil Frey for researching other interesting aspects of the rubber sector in Kerala and thus broadening this study.

My gratitude goes to Simon Pare for his proofreading and comments and to Simone Schoch, Mario and Emil for their patience and love and for accompanying and supporting me on this journey.

The thesis was carried out within the framework of the participation of the Human Geography Department, Zurich University, in the NCCR North-South Research Partnership funded by the Swiss National Science Foundation and the Swiss Agency for Development and Cooperation. Whom I thank for their financial support.

BALZ STRASSER

# Abstract

In the face of economic liberalization, a reduced role of the state, and the changing institutional setting affecting less developed countries, it has become important to understand the various impacts of these processes on the livelihoods of rural households. Empirical studies show, for example, that small holders are facing more and more difficulties in dealing with declining terms of trade and the fluctuating prices of agricultural commodities, which play an important role in the income of many small holder producers in rural areas. There is a hypothesis that, since the beginning of these processes, the opening-up of rural areas to the 'global world' has induced a shift from solely agricultural and farm income towards a more diverse income portfolio. A second hypothesis is that the local institutional setting plays a key role in supporting or hindering the diversified livelihood strategies of small holders. This study takes these as its research hypotheses and seeks to validate them through a crop- and locality-specific case study.

The Indian natural rubber sector provides good opportunities for in-depth studies; it is an example of a sector which has undergone many changes since it was integrated into the New Economic Policy of the Indian government at the beginning of the 1990s. These processes have affected rubber holders in Kerala—around one million growers cultivating an average of 0.5 ha of rubber plantation—in different ways. The case study looks at natural rubber holders located in Thalanadu Panchayat in central Kerala.

Analysis has shown the need to distinguish between different types of small holdings in order to come to any meaningful conclusions about their livelihood situations. In fact, analysis of the income-generating activities of different types of rubber holdings in Thalanadu shows that there are multiple livelihood portfolios even within the same locality and context. Thus, this study gives a new perspective on these coping processes by developing a typology of rubber holders. The extent of the *diversified status* and the type of income activities pursued depend on different factors, the most important of which are the total size of the holding and/or the access to regular wage employment in the non-agricultural sector. The most diverse portfolios are to be found at the extremes of the holding-size range, i.e. on the most marginal as well as on the largest holdings, which also represent the poorest and the richest ones. The holdings in the middle of the range are less diverse and more specialized in rubber cultivation. For marginal holdings, the key driving force towards a diverse income situation is found to be their vulnerability due to their limited monetary income. Other factors such as seasonality, climatic uncertainties, price fluctuations and a greater overall geographical marginality of these holdings also influence their decision-making, which can be best described as risk-minimizing. The key driving forces behind the diversified situation of larger holdings are their opportunities to access new activities due to investment capabilities, the availability of land resources and, in particular, their ability to take risks and cope with failure if need be. This thesis also analyses the change from one portfolio to another between 1995/6 and 2003/4, so as to study *diversification* as a process. In general, diversification was not very pronounced, since many holdings were already diverse beforehand. Thus, the study concludes that the price fluctuations in the rubber sector (the 'rubber crisis' in the mid-1990s) were not a major (or even important) reason to start or increase diversification—and this is true for all types of holdings.

The research on a range of local institutions and organizations—namely, the Rubber Board replantation subsidies, the rubber marketing institutions, the Vanilla Promotion Scheme, loans from informal and formal institutions, and social networks and social status—shows their influence in supporting or hindering the diversified livelihood strategies of natural rubber holders. The study shows that most play an important role in promoting or hindering diversified incomes. However, each institution and organization affects each type of holding in a different manner. Institutions can have a promotional effect on diversified livelihoods on some types of holdings, whilst being restrictive for others. The study thus includes a differentiated analysis of selected local institutions and organizations.

In conclusion, the findings suggest that it is important to acknowledge that there are many different types of rubber holdings, each with specific income strategies and portfolios. This diversity is, however, not reflected in the approaches of the existing extension services. The extremely heterogeneous situations of rubber holdings would, however, imply that any kind of institutional support which intends to improve the livelihood of rubber growers has to be tailored to the particular situation and the particular needs of each type of holding.

# Glossary

| | |
|---|---|
| Acre | 1 acre = 100 cents = 0.4046 ha |
| BPL | used to describe absolute poverty, i.e. people living below a defined poverty line (defined in 2003-4 as earning less than Rs. 325 per month) |
| *Chitty* | informal saving and credit scheme |
| Crore | unit of 10 million |
| *Ezhava* | a local caste |
| Hectare (ha) | 1 ha = 2.471 acres |
| *Krishi Bhavan* | local Agricultural Office of the Department of Agriculture |
| *Kudumbasree* | a women's self-help group |
| Lakh | unit of 100 thousand |
| *Malayalam* | language of Kerala |
| *Malayalees* | people of Kerala |
| Panchayat | local political-administrative unit in rural areas |
| Panchayati raj | local self-government |
| Rs. to Euro | In 2002 the Rs. was approximately 42 to the Euro. By 2005 the Rs. was 57 to the Euro.* |
| Rupee | Indian currency, abbreviated as *Re.* |
| Scheduled Caste (SC) | constitutional term for untouchable castes |
| Scheduled Tribe (ST) | constitutional term for tribal or *adivasi* people |
| *Taluk* | an administrative sub-unit of a district incorporating several revenue villages |

*According to http://oanda.com/convert/classic, accessed on 24 January 2006

| | |
|---|---|
| Toddy | an alcoholic drink fermented from palm sap |
| Ward | administrative division within a Panchayat for the purpose of administration and election |

# Abbreviations

| | |
|---|---|
| APL | person or household Above Poverty Line |
| BPL | person or household Below Poverty Line |
| CSI | Church of South India |
| GDP | Gross Domestic Product |
| IFAD | International Fund for Agricultural Development |
| NABARD | National Bank for Agriculture and Rural Development |
| NCCR North-South | National Centre of Competence in Research North-South |
| NGO | Non-Governmental Organization |
| NSDP | Net State Domestic Product |
| NSS | National Sample Survey |
| NR | Natural rubber |
| PAMS | Partnership Action for Mitigating Syndromes of Global Change (specific programme of the NCCR North-South) |
| PDS | Public Distribution System (of food and essential commodities) |
| QRs | Quantitative Restrictions |
| RB | Rubber Board of India |
| RNFE | Rural Non-Farm Economy |
| RPS | Rubber Producers' Society |
| RRII | Rubber Research Institute of India, part of the Rubber Board |
| SC/ST | Scheduled Caste/Scheduled Tribe |
| WTO | World Trade Organization |

CHAPTER 1

# Introduction

## 1.1 THE CHALLENGES FOR SMALL HOLDERS IN GLOBALIZED AGRICULTURAL MARKETS

An estimated 75 per cent of the world's poor still live in rural areas, even though urbanization is increasing at a rapid pace (Lutz et al. 1998: 1), and that a large part of them are engaged to some extent in agricultural production (IFAD 2001: 15; Crole-Rees 2002: 2; DFID et al. 2004: 3). Of these, a large proportion are dependent on agricultural commodity markets.

Although many countries have been trading in world markets for agricultural commodities for hundreds of years, the last two decades of the twentieth century have brought a wave of globalization and economic integration, which has been of utmost importance for the world's community. This was made possible by a combination of lower trade barriers and technological innovations which greatly reduced transaction costs for movements of goods, people, and capital (Bruinsma 2003: 268). As Mertz et al. state, 'each country and region has experienced these changes differently, but all have become more firmly involved in the wider process of globalization, for good and for bad' (2005: 210).

During the same period, macroeconomic difficulties and debt problems have forced many less developed countries to start reform processes with a goal of promoting growth-led development.[1] Reforms in national-level policies have led to processes of withdrawal and deregulation (less involvement of the state), while macroeconomic reorientation has brought the liberalization of agricultural and industrial commodity markets alongside reforms in other sectors (e.g., education). The consequences of these reforms have been manifold. While some scholars state that they have been instrumental in bringing countries back to sustained economic growth, there are critical voices which cite the many disadvantages and constraints of such reforms, especially for the Least Developed Countries (LDCs). Proponents of liberalization and deregulation, on the other hand, talk mainly about the positive effects linked to the creation of 'right prices' through the elimination of price distortion. They also claim that deregulation of the agricultural sector results in a more efficient allocation of resources due to the reduction of public institutional interventions in the market chain. Deregulation also provides incentives for farmers and the private sector to invest in crops with comparative advantages, under the condition that developed countries continue to dismantle their barriers to agricultural and agro-industrial trade (Binswanger and Lutz 2000). Arguments against liberalization are based mainly on the imperfect nature of world agricultural markets and the huge subsidization of agriculture in most developed countries (Clairmont and Cavanagh in Sekhar 2004: 4729). The negative consequences on farm incomes and employment have also been identified (Nayyar and Sen 1994 in Nair and Menon 2004: 7).

In many developing countries, these changes have coincided with changes in land use, the degeneration of ecological resources (deforestation, land degradation), poor hygiene and sanitation, severe diseases (e.g., HIV), low education and frequent political instability. Apparently increasing climatic variability is another important constraint.

## 1.2 THE PROBLEMATIC NATURE OF AGRICULTURAL COMMODITY MARKETS

In the rural areas of less developed countries agricultural commodity markets play an important role in the income of many small holder producers. Most of these markets, however, display specific characteristics which are known under the generic term 'commodity problem'. The 'commodity problem' is characterized by the combination of constantly declining terms of trade[2] and the high price volatility of many crops. Producers therefore face the dual problem of low returns and high risks (Page and Hewitt 2001 in DFID et al. 2004).

Analysis has shown that the 'soft' tropical agricultural commodities[3] produced by many smallholders in rural areas are particularly affected by a long-term decline in prices. This is due to the relatively inelastic demand of these markets and to the lack of differentiation among producers, leading to competitive markets. This decline in prices is expected to continue in the future because of increasing supply (e.g., due to technological improvements) and falling demand (due to slower population growth) and the development of synthetic substitutes (DFID et al. 2004: 6). In addition to their long-term decline, the prices of many agricultural commodities show a high degree of volatility. The characteristic behaviour of commodity price cycles is one of long periods of low prices with shorter periods of price spikes (Gilbert 1999 in DFID et al. 2004: 6). In his study of agricultural price volatility in international and Indian markets C.S.C. Sekhar also shows that the intra-year variability in commodity markets is even more important than the average fluctuation over many years (inter-year) (Sekhar 2004: 4734).

Empirical evidence shows that smallholders are facing more and more difficulties in dealing with the decline in terms of trade and the fluctuating prices of commodity markets (DFID et al. 2004). As Sekhar states, smallholders in a country like India, with a low propensity to save and poor access to efficient saving instruments, cannot cope with the revenue variability that results from fluctuations in output prices. According to Sekhar (2004: 4729), they do not possess the requisite know-how for crop diversification and also lack access to appropriate technology.

## 1.3 FROM FARM-INCOME TO DIVERSIFIED LIVELIHOOD 'STRATEGIES'

The thinking about the role and status of the smallholder farmer and his/her importance for rural development has greatly evolved over the last fifty years. As Frank Ellis and Stephen Biggs state, two important 'paradigm shifts' have taken place. The first shift took place in the early to mid-1960s when small farmers came to be considered as the very engine of growth and development. This new 'small farm efficiency' paradigm was a break with the dominant idea of the 1950s that the subsistence farm sector played only a passive role in the process of economic development. The 'small farm first' narrative suggests that 'agriculture plays a key role in overall economic growth by providing labour, capital, food, foreign exchange and a market in consumer goods for the nascent industrial sector in a low-income country' (Ellis and Biggs 2001 : 441). In this paradigm, both growth and equity goals appear to be satisfied simultaneously via the emphasis on small-farm agriculture. In this first phase, state-led and top-down rural development programmes were implemented with their primary focus on the agricultural sector (extension, marketing and infrastructure) and the aim of backstopping agricultural income on small farms. Rural people were looked at as being essentially farmers who depended on agricultural subsistence and cash crop production. During the 1980s and 1990s, the second main 'paradigm shift' took place. The focus switched from external technologies and national-level policies to more bottom-up, grass-roots approaches. This was the advent of farming systems research, the acknowledgment of indigenous technical knowledge and the rise of participatory methods. In many countries, NGOs took over the implementation of rural development projects. With the drawing back of the state-led development initiatives, the stage was set for the deregulation and liberalization processes described earlier (section 1.1).

Since the beginning of the processes of globalization and economic liberalization, the uncertainty of agricultural markets and the opening up of rural areas to the 'global world' have induced a shift from solely agricultural and farm income towards a more diverse income portfolio. New approaches such as the sustainable livelihood approach have been adopted to analyse and understand the income of rural households. As Ellis and Biggs state, these approaches are currently challenging the idea that rural individuals or families earn their living mainly from agriculture (Ellis and Biggs 2001: 444). In fact, empirical research suggests that income from agriculture tends actually to represent only 35-60 per cent of the total livelihood income of rural households in South Asia and sub-Saharan Africa (Reardon 1997: 737; see, for example, Lanjouw and Shariff 2004: 4433). In reality, wages and salaries from activities which have little or nothing to do with agriculture, as well as remittances and other transfers, are always important (Ellis and Biggs 2001).

The livelihood approach—the assets and diverse strategies of poor rural households—has introduced a whole body of empirical research and literature on *livelihood diversification* and the *diversification* of assets and income. As Barrett et al. state in their review article about non-farm income diversification, 'diversification is the norm' (Barrett et al. 2001: 315). Since Ellis' contributions on the relevance of rural livelihood diversification (Ellis 1998, 2000b), the role, determinants and effects of income diversification have increasingly become a topic of concern for many studies (see, for example, Barrett et al. 2001; Crole-Rees 2002; Mertz et al. 2005).

Though it has been empirically shown that many rural households have moved away from sole agricultural income, it is also argued that many others are still predominantly reliant on income from agriculture. This might be even more important in regions which depend heavily on one or a few agricultural commodities. The focus on *diversification* has certainly corrected the idea that rural households are dependent only on agricultural income. However, the dominant perception seems to be swinging towards the opposite, i.e. that income from the agriculture sector is no longer so important. It is argued here that one has to look in detail at the income distribution within households and assess not only the aggregate importance of non-farm income, but also which types of households do diversify, to what extent and why. And also, why others do not want to—or cannot—diversify, and prefer or have to stay dependent on agricultural income only.

## 1.4 THE ROLE OF THE LOCAL INSTITUTIONAL AND ORGANIZATIONAL SETTING

Institutions and organizations play a key role in development. Over the last decades, a whole body of literature in different development research disciplines has analyzed institutions and organizations (see, for example, Douglas 1987; North 1990; Scott 1995). Institutions have been defined in very different ways. In this work, institutions are defined, following Ellis (2000b: 38), as the formal rules, conventions and informal codes of behaviour that comprise constraints on human interactions, while organizations are groups of individuals bound by the common purpose of achieving certain objectives (North 1990: 5).

In recent years, trends have emerged towards changes in the institutional setting, for example, market-led growth, trade liberalization, privatization and the reduced role of the state in economic development, alongside organizational reforms in many developing countries. Within this context, it has become important to understand better the manifold ways in which rural households react to a changing institutional and organizational setting and how the linkages to this setting affect their livelihood strategies. The fact that the institutional environment is often used as a 'black box' in many projects makes this doubly important

(Nuijten 1999: no pagination). As Nuijten states, it is given a central role, but its content remains 'spectacularly unclear' (ibid.: no pagination). In cases where institutions and organizations are defined more explicitly, the focus tends to be placed on formal ones (e.g., credit unions, cooperatives) because they are the most 'visible'. More informal ways of being organized are much harder to study and to understand. To close this gap, it is essential to study the impact of institutions only on a limited geographical area. This was done in this study by examining the concept of *local institutional and organizational setting,* which refers to the specific manifestations of institutions and organizations in the geographical area under study, even though institutions may cross the boundaries of this area (Uphoff 1993: 608-9).

## 1.5 ISSUES WITHIN THE INDIAN NATURAL RUBBER SECTOR

The Indian natural rubber (NR) [4] sector offers a good opportunity for an in-depth study since it is an example of a sector which has undergone many changes since macro-policy reforms were initiated in 1992. The high world market price in the mid-1990s, together with (re)plantation incentives offered by the government, led to increased planting of rubber trees; the increased output a few years later fell at precisely the time when the price collapsed (George 1999: 190; Patnaik 2002: 131; Rubber Board 2003 a). The price collapse affected the whole sector. However, growers with plantations in marginal agro-climatic zones with a lower average yield were affected more than planters in traditional rubber areas.

The hypothesis of this study is that rubber holders in marginal areas cope with these changes through a diversification of their income-generating strategies. A second hypothesis is that the local institutional setting plays a key role in hindering and/or supporting the coping and income-generating strategies of these small holders.

These hypotheses were compared with a case study based in the state of Kerala, south India. Field research was conducted in central Kerala, one of the main growing areas for natural rubber. Thalanadu Panchayat in Kottayam District was chosen as a study location (see Fig. 3.2). Thalanadu was chosen because it lies in the higher midlands of central Kerala, which is a marginal rubber area as the agro-climatic conditions are less favourable for the cultivation of rubber.

## 1.6 RESEARCH OBJECTIVES AND QUESTIONS

The first objective of this study is to examine the income-generating strategies of marginal rubber smallholders in Thalanadu Panchayat in central Kerala during the years 2003-4. The focus will be on the changes in the income-generating strategies that have appeared since 1995-6, so as to understand the influence of the fluctuating natural rubber prices on the livelihood strategies of rubber holders. As a second objective, the study examines the local institutional setting (including both formal and informal institutions) and its influence on the income-generating strategies of rubber holders.

Thus, the overall objective of the study is to understand in detail the different livelihood strategies of the different types of rubber holdings, the importance of diversification within these strategies and their links to the local institutional and organizational setting. Empirical findings are expected to provide valuable information which will be useful in formulating sound and appropriate policies to tackle constraints and improve opportunities, especially for the smallest rubber holders.

The following are the specific research questions of the study:

- Which income-based typology of rubber holdings exists in the study area?
- What were the income-generating strategies of the different types of rubber holdings in the study area in the year 2003-4?
- How important were the diversified livelihoods in 2003-4?
- How have income-generating strategies of the different types of holdings changed during the 'rubber crisis' (1997-8 to 2001-2)? What was the role of livelihood diversification during that period?
- Which local institutions and organizations are of importance for rubber holders?
- What is the role of the local institutional and organizational setting in supporting and/or hindering the income-generating strategies of the different types of rubber holders?

## 1.7 OUTLINE OF THE STUDY

The work is divided into eight chapters. After the introductory chapter, chapter 2 gives the theoretical and conceptual background of the study, as well as an analytical framework. Chapter 3 describes the research framework and the methodology used for the empirical research. Chapter

4 introduces the natural rubber sector in Kerala and thus gives the socio-economic context of the research location. The subsequent chapters present the results of the case study. Chapter 5 gives the results of the survey and an overview of rubber holders and their characteristics in Thalanadu, while chapter 6 first describes the local rubber cultivating practices, before suggesting a typology of rubber holders in Thalanadu. It goes on to analyse the income-generating and diversification strategies of each type of holding. The last part of chapter 6 describes the impact of the 'rubber crisis' in the second half of the 1990s. Chapter 7 first presents the local institutional and organizational setting in the study location before describing how selected institutions and organizations hinder or favour diversification. Chapter 8 concludes the study by presenting the main findings and the theoretical and conceptual implications. The last part of chapter 8 is devoted to the possible implications for policy and research.

## NOTES

1. The structural adjustment programmes designed, implemented and sustained both by the World Bank (WB) and the International Monetary Fund (IMF) are a good example of these processes.
2. Declining terms of trade refers to commodity prices rising less rapidly than those of manufactured goods.
3. 'Soft' refers to commodities which are mainly produced in developing countries and consumed in developed (e.g., coffee, cocoa and tea), while 'hard' refers to commodities for which market intervention is significant in developed countries (e.g., sugar, cotton, wheat, rice) (DFID et al. 2004: 6).
4. This work is essentially about natural rubber extracted from the latex collected from the *Hevea* tree. Therefore, *natural rubber* and *rubber* will be used synonymously and abbreviated as NR. If at times *synthetic rubber* is meant, it will be specifically mentioned as thus.

# CHAPTER 2

# Theoretical and Conceptual Background

This chapter gives the theoretical background to the different concepts used in this work. Section 2.1 embeds smallholder farmers in past and contemporary rural development theories and debates, discussing the role of agriculture in general and smallholders in particular in rural development. Section 2.2 details the rural livelihood diversification debate and its conceptual issues. Section 2.3 presents the main determinants of diversification. Section 2.4 looks at the theoretical debates concerning the role of the local institutional and organizational setting in livelihood diversification. A description of the possible effects and outcomes of livelihood diversification, as well as the typologies of diversification strategies, is given in section 2.5. Finally, section 2.6 presents the analytical framework of the study.

## 2.1 SMALL HOLDERS AND RURAL DEVELOPMENT

Different paradigms and bodies of thoughts have marked the landscape of rural development thinking. This section describes the main theories since the 1950s, so as to put the debate on rural livelihood diversification that first arose in the 1980s into context.

### 2.1.1 Rural Development Theories in the Past

*Dual-economy theory*

The dominant theoretical thinking during the 1950s was the dual-economy theory (Boeke 1953; Sieberg 1999), which refers to an economy in which rich, capital-intensive modern sectors exist in the same model as comparatively poor, traditional, labour-intensive sectors. The subject of debate was whether an economy should achieve economic growth through its technically advanced sectors or whether resources should be spread evenly across the whole economy to achieve a more balanced growth. According to this theory, the subsistence sector was not expected to raise productivity or lead to growth and was therefore assessed as playing a passive role in economic development. Instead, the modern sector of large-scale 'modern' agriculture (plantations, estates, commercial farms) and the manufacturing industry were expected to stimulate economic growth. The development trajectory foresaw a move from agrarian to industrial-based national economies (Bryceson 2000a: 20). The following quotation from Streeten (1972 in Bryceson 2000a: 20) gives a clear idea of the conceptualization and the role of smallholders in economic development at the time:

> A widespread view is that agricultural progress is necessary in order to supply, firstly, *a surplus of labour* for industry, secondly, *a marketable surplus of food* for industrial workers and, thirdly, *an investible surplus of savings* for urban industry. [emphasis by Streeten]

On the agricultural front, one underlying argument in favour of this theoretical proposition was the existence of economies of scale, i.e. that large farms could use their resources and modern technologies more efficiently than small farms. As Ellis and Biggs (2001: 440) suggest, this thinking was also important in socialist strategies of agricultural development as practised in the former Soviet Union and in low-income developing countries that had socialist-leaning governments in the 1960s and 1970s.

### *Small-farm first model*

In the first half of the 1960s, thinking switched from the dual-economy theory towards the idea that the so-called traditional or subsistence agriculturalists in low-income countries could form the basis of agriculture-led processes of economic development. Ellis and Biggs (2001) call it the 'agricultural growth based on small-farm efficiency' paradigm or in short 'small-farm first', while other scholars write of the 'small-farm model' (Ashley and Maxwell 2001: 406). The dominant paradigm of the 'small-farm first' idea is that the rural agricultural sector played a key role in overall economic development, though it was commonly accepted that agriculture would decline in its share of GDP as development progressed (Ellis 2000b). The agricultural sector was needed to provide labour, capital, food and foreign exchange, as well as a market for consumer goods for the nascent industrial sector in a low-income country. Rising agricultural output and incomes were seen as a prerequisite for and a stimulant of non-farm growth. As Singh (1990 in Ellis 2000b: 22) puts it, 'the growth of the non-farm economy depends on the vitality of the farm economy; without agricultural growth in the rural areas, redressing poverty is an impossible task'. Small farms were seen as being efficient, not least in exploiting new green revolution technologies which were largely scale-neutral. The 'small-farm first' model is interlinked with much academic work of the 1960s which emphasized the benefits of agricultural growth on small farms and was not restricted just to the commercial farm sector. The following are some important contributions (based on Ellis 2000b: 21-2; Ashley and Maxwell 2001):

- Small farmers are rational economic agents and make efficient farm decisions;
- They maximize returns on land because (a) they use family labour intensively and in doing so avoid the constraint of having to supervise a large, hired labour-force, and (b) they tend to be located in places which are difficult to reach with mechanization;
- They innovate as successfully as large farmers because new technologies (e.g., high yielding varieties and then required input combination) are scale-neutral;
- There is an 'inverse relationship' between farm size and economic efficiency, such that small farmers are more efficient than large ones (Berry and Cline, in Ellis and Biggs 2001);
- Rising agricultural output in the small-farm sector results in 'rural growth linkages' which spur the growth of labour-intensive non-farm activities in rural areas, and these are higher than for large farms (Mellor 1976, in Ellis and Biggs 2001). This idea has been particularly lasting.

An important attribute of the 'small-farm first' model is that both growth and equity goals appear to be satisfied simultaneously, leading to a win-win situation. This combination of outcomes is stated as being one reason for the 'enduring success' of this narrative that subsisted well into the 1990s (Ellis and Biggs 2001: 441).

*First approach to rural development*

The term *rural development* came into widespread usage in the mid-1970s following the publication of a sector policy paper on rural development by the World Bank. The term was primarily associated with the empirical evidence and acknowledgment that the majority of the population in developing countries with incomes below a stated poverty line resides in rural rather than in urban areas (World Bank 1975: 4-5). In its early years, the emphasis of rural development was on raising the output and incomes of small, poor farmers; this was in line with the 'small-farm first' paradigm. Consequently, donors and governments supported small-farm agriculture with research and extension programmes, irrigation schemes, credit facilities, agricultural inputs and so on. Indirectly, small farms were supported with investments in rural infrastructure such as roads, transport, grain stores and other marketing facilities. As Ellis notes, 'while other aspects of rural welfare such as education, health services, water supplies and so on were also accorded some weight in government and donor plans and projects, there is no doubt that small-farm output growth was writ large as the dominant strategy for improving rural welfare' (Ellis 2000b: 26). The consequence was that, until the 1980s, mainstream development thinking neglected the livelihood strategies of the rural poor, for whom subsistence farming was unable to provide sufficient income possibilities. Starting from the mid-1980s, this lack of attention was slowly redressed with a new paradigm shift within rural development.

*'Process approach' to rural development*

One of the main lines of critiques, arising in the mid-1980s, concerned the role of government as the sole and predominant agent capable of effecting rural development. It was argued that the state-led and top-down approaches of rural development projects were limited in their potential to address diverse problems in the rural world and that rural development should instead be looked at as a 'participatory process that empowers rural dwellers to take control of their own priorities for change'. This criticism of 'big government' with its national-level policies was fuelled by the structural adjustment and market liberalization agenda of the early 1980s; these aimed at making governments disengage from their previous large-scale 'management' of the agricultural sector (Ellis

2000b: 26; Ellis and Biggs 2001: 443). The same period saw the emergence of grass-roots and bottom-up approaches, although these came from different sources. The following list contains some of their key strands (based upon Ellis 2000b; Ellis and Biggs 2001: 443):

- Farming systems research based on the growing argument that the Green Revolution might not necessarily work in diverse, risk-prone and resource poor environment (Byerlee et al. 1982) as it did in monocrop farming systems;
- Indigenous technical knowledge (ITK) and the growing acknowledgement that the poor were able to contribute to solutions to their own problems themselves (Richards 1985);
- Participatory methods, starting from rapid rural appraisal (RRA) and evolving into participatory rural appraisal (PRA) and participatory learning and action (PLA) (Chambers 1980);
- 'Actor-oriented' approaches to rural policies and projects, emphasizing that participants in rural development (including the poor themselves) are actors with differing understandings of the process of change in which they are involved (Long and Long 1992).

Linked to these strands was the rise of non-governmental organizations (NGOs) as important vehicles for implementing rural development projects, the rise of gender as a concern in rural development, the rejection of overarching theories as a guide of action (Booth 1994) and, as mentioned above, the increasing impact of structural adjustment and market liberalization.[1] As summarized by Ellis, 'The shift is from the general to the particular, from seeking single solutions with widespread application to addressing specific problems in a limited context, from implementing solutions from above to permitting solutions to be generated from below' (2000b: 27).

### 2.1.2 Small holders in the Contemporary Rural Development Debate

In their review of contemporary rural development, Ashley and Maxwell state that 'trends and discontinuities in the character of rural areas generate a rural development *problematique* sharply different from that of the past' [emphasis added] (Ashley and Maxwell 2001: 397). They point first to the fact that the term 'rural' is often used quite ambiguously (ibid.). A study by IFAD (International Fund for Agricultural Development) defines rural people, firstly, as usually living in farmsteads or settlements of five to ten thousand people separated by farmland, pasture, trees or scrubland, and, secondly, as spending most of their working time on farms. However, national distinctions between rural and urban are arbitrary and varied (IFAD 2001: 17). Others define 'rural' as being characterized by high transaction costs,[2] associated with long distances and poor infrastructure, or by geographical conditions which increase political transaction costs (Binswanger and Deininger 1997). Nevertheless, Ashley and Maxwell make an analysis of long-term rural development indicators and projections for selected developing countries. Accordingly, they suggest that 'in all its diversity, rural space has changed fast and will change faster' (Ashley and Maxwell 2001: 400-1). The following are some selected key points from their study:

- Rural population will begin to stabilize, but with a changing dependency ratio;[3]
- Connectedness of rural areas will improve with more roads and other infrastructure (e.g., for communication);
- Human capabilities will improve with better education and health;
- The great majority of rural people will be functionally landless, either without land altogether, or with only a small homestead plot;
- Most rural income in most places will be non-agricultural in origin (though with linkages to agriculture in many cases);
- Farms (other than part-time subsistence or homestead plots) will be larger then at present, and grow even larger;
- Most farms will be predominantly commercial, i.e. buying most inputs and selling most of their output, and marketing systems will be integrated;
- As a result, agriculture's contribution to the GDP will be no more than 10 per cent, and agriculture will not contribute more than 10 per cent to exports (with perhaps more in Latin America and sub-Saharan Africa);
- Agriculture will become a net recipient of government revenue.

For the rural population, these patterns of change do not guarantee any particular outcome. However, the change does emphasize the need to understand the diversity of rural areas and the complexity of livelihoods and livelihood strategies (ibid.: 401).

In contemporary research about rural development, the discussion of whether agriculture is the best way to reduce rural poverty is still *en vogue*. In their study, Irz et al. show twelve separate reasons why agricultural growth might be expected to reduce poverty, at farm level, in the rural economy, and nationally. However, they also provide a long list of necessary conditions for this to be the case, and repeatedly stress that definitive outcomes cannot be predicted a priori. However, reviewing empirical evidence, they conclude that the benefits are substantial: 'for the poor, extra farm jobs and higher wages may be the single most obvious benefit [of agricultural growth]; followed by the impact of additional spending in the rural economy; and the value to the national economy and social welfare of reduced costs of food' (Irz et al. 2001, in Ashley and Maxwell 2001: 402).

This seems to confirm a positive relationship between agricultural growth and poverty reduction, as other studies have also shown (see Ashley and Maxwell 2001).[4] Ashley and Maxwell (ibid.: 403-4), however, state four reasons which question these findings. Firstly, many of the expected positive effects of agricultural growth depend on small farms being at the forefront, but this may be a problematic assumption. Secondly, the long-term global fall in agricultural commodity prices and terms of trade has undermined the profitability of agriculture as a business. Thirdly, agriculture is pushing against natural resource boundaries, particularly soil and water. Finally, it seems to be a fact that diversification out of agriculture is already occurring in many dynamic rural economies. Other arguments are linked to the above points: falling or stagnating prices require low margins (i.e. lower income for farmers) or aggressive technical or institutional change to lower costs, or a shift to new, higher-value products (e.g., export crops). Furthermore, in the competitive environment of world markets, one has to be faster than one's competitors (ibid.: 404; see also World Bank 2003). Kydd and Dorward (2001: 472) put forward the argument that governments in developing countries have to invest more in public goods for agriculture and need to find ways to reduce costs and increase efficiency. Opposition to parastatal involvement in agricultural input and credit supply results from inadequate understanding of the nature and importance of *transaction costs* within smallholder agriculture. Furthermore, 'the (micro-economic) institutional arrangements and transaction costs and risks in output markets, input delivery and seasonal finance need to be given as much attention as has traditionally been given to the micro-economics of production' (Kydd and Dorward 2001: 475). Since it is difficult to predict how governments of developing countries will approach these questions, the issue remains whether small farmers are financially and socially profitable, now and in the long-term, with or without additional investments in rural public goods.

There is nowadays much evidence in favour of the 'small-farm first' model, the main one being the ability of farmers to make efficient decisions, use family labour intensively, maximize return on land and be successful innovators because new technology proves to be scale-neutral and no more risky than the traditional one. A study by IFAD (2001: 79) validates some of these points: in some less developed countries (Colombia, Brazil, India, Malaysia), the land productivity of small farms is at least twice that of the largest ones. Generally, this results from higher employment opportunities: 'usually, small farmers' advantages are due less to high yields of the same crop than to a higher value crop mix, more double-cropping and inter-cropping, and less fallowing' (ibid.). However, due to the changes in rural areas and to the unclear outcomes and consequences of agricultural growth, the potential role of small farmers has to be re-evaluated. As Ashley and Maxwell mention (2001: 407), 'there is a case against [small farmers], particularly, with regard to small farms in the new style rural spaces . . . characterized by much greater income diversity, stronger urban-rural links, and greater integration into the world economy'. They put forward the following points in favour of their argument:

- Land is often not a scarce resource on small farms (especially, but not only, for part-time farms), where cash or seasonal labour may be limited;
- Part-time farmers may not see the need to maximize the return from farming;
- Small farmers are more likely to grow low-value staples for self-sufficiency;
- New technology is capital-biased, reflecting long-term investment by the private and public sectors in meeting the needs of farmers in the north;

- The skills required to manage new technologies are beyond the scope of many small farmers;
- Small farms pay more for inputs and receive less for outputs than large farms;
- New commodity chains impose quality and timeliness requirements which small farmers find hard to meet (and which cooperatives cannot help with).

However, empirical evidence in favour of these arguments is fragmentary. Killick, reviewing evidence of the impact of globalization on the rural poor, concludes that he is concerned about 'the long-term ability of many poor smallholder farmers to respond adequately to population pressures, growing international competition and agricultural commercialization' (Killick 2001: 175). An important point for the future of small farms is whether they can enhance their production efficiency to be able to compete with large-scale agricultural farms. Kydd and Dorward point out that the efficiency of smallholder farming may be 'breaking down, where globalization intrudes, non-traditional crops are promoted, and agricultural modernization involves increasing use of capital' (Kydd and Dorward 2001: 471). One possible solution is the payment of poverty- or environment-related social subsidies to small farms.[5]

### 2.1.3 The Rise of the Rural Non-Farm Economy

Numerous recent studies on the rural sector in developing countries have pointed to the growing importance of the so-called rural non-farm economy.[6] In fact, recent surveys suggest that non-farm income sources account for 40-5 per cent of the average rural household income in sub-Saharan Africa (Reardon 1997; Start 2001) and 30-40 per cent in Asia (Lanjouw and Shariff 2004: 4433), with the majority coming from local rural sources rather than from urban migration. Some of this, perhaps much of it, is of course directly or indirectly related to agriculture (Ashley and Maxwell 2001: 408). There is, however, also a growing part of the rural non-farm economy that is independent of agriculture, as is shown by a study in Ghana (Canagarajah et al. 2001, in Ashley and Maxwell 2001: 408).

In order to reconnect with the different rural development paradigms elaborated above, one can state that the emerging evidence that the rural poor often depend on non-farm (and sometimes non-rural) sources of income in order to sustain their livelihoods is a key issue. Ellis (1998) states that this fact could contribute to a reappraisal of the validity of the 'small-farm first' orthodoxy. Furthermore, Ellis and Biggs mention that the so-called *sustainable livelihood approach* (Carney 1998; Scoones 1998) could provide a challenge to the 'small-farm first' orthodoxy. They point to the interesting fact that the approach does not require the poor rural individual or family to be a 'small farmer'. The livelihoods concept takes an open-ended view of the combination of assets and activities that turn out to constitute a viable livelihood strategy for the rural family. As they go on to say, the starting point of the livelihood approach — the assets and diverse strategies of a poor household—is thereby fundamentally different from the principles underlying 'small-farm first' thinking, and can lead analysis in new directions (Ellis and Biggs 2001: 444-5).

## 2.2 RURAL LIVELIHOOD DIVERSIFICATION

### 2.2.1 The Rise of Livelihood and Diversification Studies[7]

The roots of livelihood studies can be found in farming systems research and household studies. Household studies appeared in the 1980s, when a more actor-oriented perspective was adopted in development studies. This perspective still emphasized inequalities in the distribution of assets and power, but it also recognized that 'people make their own history' and that economic concerns are not necessarily of primary importance (de Haan and Zoomers 2005: 28). This new perspective focused on the world of lived experience, 'the micro-world of family, networks and community' (Johnston 1993 in de Haan and Zoomers 2005: 28), and it drew attention to related issues such as poverty, vulnerability and marginalization. A micro-orientation became dominant, with a clear focus on households.[8]

In various types of household studies, the focus on *household strategies* received increased attention as a means of capturing the behaviour of low-income people. The concept of household strategies highlighted the active or even proactive role played by the poor, which was in contrast to earlier tendencies to perceive poor people as passive victims. Household studies made it possible to explore differing responses to general structural conditions, as well as to analyse changes specific to subgroups of the population. Many of the household studies conducted in the 1980s appeared under the heading of *New Household Economics,* focusing on labour and land allocation and income strategies and using microeconomic household modelling as an explanatory tool. Subsequent studies of households used a variety of concepts, of which the most common were *survival strategies*, although Long (1984, in de Haan and Zoomers 2005: 29) was already calling them *livelihood strategies*. Studies on survival strategies were more sociologically than economically inspired and were mainly interested in the micro-social behaviour of poor people in coping with and surviving different types of crises, such as falls in prices, droughts and famines. Many household studies ended with rather pessimistic conclusions, showing how poor households were increasingly excluded from the benefits of economic growth and thus marginalized (de Haan and Zoomers 2005: 29).

In the early 1990s, a new generation of more optimistic household studies appeared, which approached households from a *livelihoods* perspective and showed how people make their living. As de Haan and Zoomers (29-30) suggest, the livelihood approach is a direct response to the disappointing results of former approaches in devising policies to alleviate poverty, such as those based on income, consumption criteria or basic needs. In Appendini's words, the central objective of the livelihood approach was 'to search for more effective methods to support people and communities in ways that are more meaningful to their daily lives and needs, as opposed to ready-made interventionist instruments' (2001, in de Haan and Zoomers 2005: 30). In the 1990s, the livelihood approach was further developed and linked to central issues of the development debate, such as food security, natural resources, environment entitlements, access and institutions. The notion of *diversification of livelihoods* came up during this period, with significant contributions from the Overseas Development Group of the University of East Anglia lead by Frank Ellis (1998, 1999, 2000b). At the end of the 1990s, Ellis wrote an excellent review of the literature on *diversification*. One of his initial statements was that the idea of diversification stands in contrast to the accepted notions of sectoral differentiation (agriculture vs. industry) and specialization (i.e. division of labour), which orthodox views of processes of economic change take to be essential for the transformation of economies (Ellis 1998: 1-2). Some scholars stated that diversification was only a transient phenomenon of people struggling for survival (Saith 1992, in Ellis 1998: 2). However, studies of the livelihoods of poor people have shown that diversification is much more than a transient phenomenon; non-farm income accounts in average for as much as 30-45 per cent of total farm income in Africa and Asia, as already mentioned.

In recent years, much work has focused on specific aspects of diversification. While some have looked at the determinants and the contribution of diversification (Abdulai and Crole-Rees 2001; Crole-Rees 2002), others assess its consequences and outcomes (Orr and Mwale 2001; Xia and Simmons 2004) and the changing livelihood strategies (Koczberski and Curry 2005; Wadley and Mertz 2005), or its diversification at a sectoral level (Joshi et al. 2004). Bryceson tries to bridge between income diversification and the broader process of 'de-peasantization' and 'de-agrarianization' (Bryceson 1999, 2000b).[9] A recent joint FAO-World Bank study on the reduction of hunger and poverty highlights the importance of diversification of production and off-farm income, both in agriculture and in non-agriculture (Dixon et al. 2001: 13).

While some of these studies are strictly economic, others are more sociological in their approach. A good review of the concepts, dynamics and policy implications appeared in a 2001 special issue of *Food Policy* (Barrett et al. 2001). Looking at all the outcomes of these studies, Ellis is right in mentioning that 'the causes and consequences of diversification are differentiated in practice by location, assets, income, opportunity and social relations; and . . . these manifest themselves in different ways under different circumstances' (Ellis 1998: 3).

Diversification studies can take place both at a household level (diversification of households) or at an aggregated level, for example, national, regional or village level (diversification of economies). The first level is best studied by analysing the livelihood strategies of the 'multi-functional' households (Ellis and Biggs 2001), the second requires a broad, and complementary, approach to long-term structural transformations (Start 2001). Furthermore, some studies do connect household-level diversification and the development of the rural non-farm economy (Delgado and Siamwalla 1997).

### 2.2.2 Defining Livelihood Diversification

The many debates around the definition of *livelihood diversification* make it necessary to devote a section to definitional aspects.[10]

In his review article, Ellis defined *livelihood diversification* as 'the process by which rural families construct a diverse portfolio of activities and social support capabilities in their struggle for survival and in order to improve their standards of living' (1998: 4). A livelihood therefore comprises not only income, both cash and kind, but also the social institutions (kin, family, compound, village, etc.), gender relations and property rights required to support and sustain a given standard of living. For example, social and kinship networks are important in facilitating and sustaining diverse income portfolios; and social institutions are critical to interpret the constraints and options of individuals and families distinguished by gender, income, wealth, access and assets. A livelihood also includes access to—and benefit from—social and public services provided by the state such as education, health services, roads, water supplies and so on (from various authors in Ellis 1998: 4).

Several components of the definition need to be elaborated. Some scholars have pointed out that *livelihood* is more than just *income.*[11] As de Haan has put it:

> Livelihood is not necessarily the same as having a job and does not necessarily even have anything to do with working. Moreover, although obtaining a monetary income is an important part of livelihood, it is not the only aspect that matters. It is quite conceivable for somebody with a low monetary income to be better off than someone with a higher monetary income. (de Haan 2000: 343)

Livelihood diversification is therefore not synonymous with *income diversification.*[12] Nevertheless, many economic studies of diversification focus on different income sources and their relation to income levels, income distribution, assets, farm output and other variables. Furthermore, there is an important difference between *income diversity* and *income diversification*. While the former refers to the composition of household incomes at a given instant in time (like a photograph), the latter interprets this as an active social process whereby households engage in increasingly complicated portfolios of activities over time (like a film). However, this conceptual difference is often not taken into account. As a matter of fact, many studies which aim to study *diversification* instead study *diversity*, i.e. the status at a given time, and not the process leading up to it. The reason for this is often a lack of comparable evidence across intervals of time which makes it impossible to study whether household livelihoods are more diverse today than they were in the past. Nevertheless, Ellis

(1998: 5) writes that 'there seems to be an informed consensus that diversity has been increasing in recent history'.

The present work is based on Ellis' broader definition of livelihood diversification as referred to above. However, part of the work also specifically focuses on income diversification as a graspable measure of diversification. This work also makes a specific differentiation between income diversification and income diversity (section 6.7.1).

In their review about non-farm income diversification, Barrett et al. take a more economic approach towards diversification: 'diversification patterns reflect individuals' voluntary exchange of assets and their allocation of assets across various activities so as to achieve an optimal balance between expected returns and exposure conditional on the constraints they face' (Barrett et al. 2001: 316). Though their definition is restricted to the diversification of income, their conceptual approach based on specific variables of interest (income, assets and activities) and their clear distinction between sectoral and spatial lines (farm, non-farm, etc.), is helpful when studying diversification empirically (discussed below).

One still needs to tackle the issue of the household being taken as the unit of analysis in diversification studies. Much has been written in development research about the concept of *household* and the underlying difficulties of defining it and thus using it as a category of analysis (Netting 1993; Ellis 2000b: 18-20; Thüler 2002; Kaspar and Kollmair 2006). With reference to Wilson (2004: 1), a household is defined in this study as a unit of production, consumption and reproduction as well as the basic economic decision-making unit. Household implies the *extended household*, which means that the household is not limited spatially and can include household members working in urban centres or abroad and sending remittances back to the main residential unit of the household, thereby contributing to the household livelihood. Therefore, all members contributing to an income (in kind or in cash) through production and consuming assets of that household (at least partially) are part of it. As Ellis mentions (1998: 6), there may be instances when this definition (which is based on a more economic approach) fails to capture important attributes of individual and collective welfare, including, for example, 'spheres of individual decision-making, power relations within the social unit, and constraints on permissible courses of action by gender'. Furthermore, it can miss aspects such as intra-household inequalities or variations in household composition concerning gender, age, and kinship. Harriss is certainly right on this point when he comes to the conclusion that economic models must always be confronted by social reality (2001, in White 2001: 7).

### 2.2.3 Variables of Interest in Diversification

Authors of livelihood studies need to have conceptual clarity about how to define and deal with the term *income*. The following discussion follows Barrett et al. (2001) who describe two main conceptual issues. First of all, the interesting variables of each study are defined. Then, they are classified following a standardized procedure to allow comparison across studies.

*Definitional aspects of income, assets and activities*

The prevailing practice in diversification studies is to emphasize the measure of *income*. Barrett et al. suggest, however, that income—though it offers a measure of direct interest because it can be clearly interpreted as a welfare outcome—does not give us any further information, such as whether it is derived as a result of a forced choice or from an income opportunity. Thus, a household's *assets* and their allocation can be a complementary measure in the study of income as part of a livelihood strategy. Assets are both a store of wealth (e.g., financial capital) and a source of income (e.g., human capital). One problem with assets is that they are difficult to value (Barrett et al. 2001: 318). Finally, one could study the *activities* of a household. Activities can be defined as services which use assets and normally lead to a flow of income. But to study only activities would miss the income generated from non-productive assets[13]—and, again, activities are very difficult to value. Since none of the three variables are unambiguously better than the others, Barrett et al. advocate 'the use of multiple indicators as cross checks on inference based on any single one' (ibid.). The present study follows this approach, tying observations of one sort of indicator (e.g., gross income) to data on another indicator (e.g., change in main income activity).

*Classification of income, assets and activities*

One source of confusion in the literature derives from the inconsistent terminology used for the classification of income and activities. In fact, terms like 'off-farm', 'non-farm', 'non-agricultural', etc., are often used as synonyms. Barrett et al. suggest a three-way classification of earned income according to *sector* (e.g., farm vs. non-farm), *function* (wage vs. self-employment) and *space* (local vs. migratory), as is shown in Table 2.1. The most basic classification follows the well-known *sectoral* distinction that is also used in national accounting systems: [14] primary sector (agriculture, mining and other extractive activities), secondary sector (manufacturing) and tertiary sector (services). The distinction between 'agricultural' (or 'farm') income and 'non-agricultural' (or 'non-farm') income follows from this. 'Agricultural' income is derived from the production or gathering of unprocessed crops or livestock, forest or fish products from natural resources; 'non-agricultural' income is derived from all other sources including processing, transport or trading of unprocessed agricultural, forest and fish products.

TABLE 2.1: ASSETS, ACTIVITIES AND INCOMES

| sector → | *agricultural* = all activities in the agricultural sector, regardless of location or function | | *non-agricultural* = all activities outside the agricultural sector, regardless of location or function | |
|---|---|---|---|---|
| function → space ↓ | *wage-employment* | *self-employment* | *wage-employment* | *self-employment* |
| *on-farm* = all activities in one's own property, regardless of sector or function (almost always self-employment) | agricultural activity, done on-farm, with wage | agricultural activity, done on farm, with self-employment | non-agricultural activity, done on-farm, with wage | non-agricultural activity, done on-farm, with self-employment |
| *off-farm* = all activities away from one's own property, regardless of sector or function | agricultural activity, done off-farm, with wage | agricultural activity, done off-farm, with self-employment | non-agricultural activity, done off-farm, with wage | non-agricultural activity, done off-farm, with self-employment |

*Source*: Author's figure, based on Barrett et al. (2001: 319).

This means that the sectoral farm/non-farm (or agricultural/non-agricultural) distinction concerns only the nature of the product and the factors used in the production process. It does not matter where the activity takes place (in the home or abroad), or whether the participants earn profit or labour income (wages or salary).[15]

The second distinction concerns the *function*. Activities and income can be divided into 'wage-employment' or 'self-employment' though there is a grey area whereby some activities could be classified as either. Barrett et al. write that a distinction between these two can be of interest because 'labour market opportunities vary enormously between poorly compensated unskilled wage labour in any sector, well compensated, generally more dependable skilled wage or salary labour . . . and self-employment in skilled or unskilled trades or commerce in the non-farm sectors or in farming' (2001: 319).

The last categories concern the *spatial* dimension. An activity can be either (a) 'local' with the two subcategories 'at-home'—or the more ambiguous term 'on-farm'—and 'local away-from-home' or (b) 'distant away-from-home' (sometimes called migratory) with the subcategories 'domestic rural', 'domestic urban' or 'foreign'.

## 2.3 DETERMINANTS OF DIVERSIFICATION

This section analyses the driving forces (or determinants) of diversification. The approach most often found in the literature about diversification is based on an economic analysis of households. It is, however, clear that social and familial constraints also apply: it is not only what people do but ultimately also their capabilities to change what they do that is influenced by their social context. This 'social embeddedness' (Rogaly 1997, in Ellis 1998: 11) must be kept in mind when interpreting economic propositions.

The driving forces analysed and linked to the diversification debate are as follows: seasonality; risk and risk management strategies; vulnerability, risk strategies; coping behaviour and adaptation; and market (financial, labour, land) failures.

### 2.3.1 Seasonality

Every rural household is confronted with seasonality as an inherent feature of its livelihood.[16] Rain, its duration, the length of the growing season, temperature variations across the year and so on, all affect the production cycle of crop and livestock—and thus ultimately the income of the farm enterprise. These seasonal factors apply both to landless rural families that depend on agricultural labour markets as they do to farm families. But seasonality also affects agricultural supply and output services such as fertilizer delivery and crop marketing. For example, trading activities (e.g., with coffee) are particularly cyclical along seasonal lines, and for some crops the trading season is very short. Economically, seasonality means that returns to labour time (i.e. income which can be earned per day or week worked) vary during the year in both on-farm and off-farm labour markets (Ellis 2000b: 58). On-farm returns are different in periods of peak labour, such as sowing or harvesting time, from periods when little or no activity can be undertaken on the farm. Off-farm returns are different when there are temporary labour markets, as, for example, to harvest a grain or tree crop, or to move recently harvested produce from farms into distribution centres or stores. Thus, seasonal changes in occupation occur as labour time is switched from lower to higher return activities (Alderman and Sahn 1989, in Ellis 1998: 11). Conceptually, it is important to separate seasonality from risk, though in practice they are closely related. In fact, seasonality can happen without necessarily involving risk. As Ellis states, 'diversification that obeys different opportunities in different seasonal labour markets does not require risk as an explanatory argument for its occurrence' (Ellis 1998: 11).

For a household, seasonality signifies that continuous household consumption does not necessarily match up with uneven income flows. If one were to leave aside risks and market imperfections, this would not be a problem per se, as long as the total income was sufficient to cover annual consumption requirements. With savings, crop storage and output sales, the unstable income can be transformed into stable consumption. However, in practice, both risk and market imperfections do exist in the rural economy (Alderman and Paxson 1992). Thus, income instability and consumption smoothening are real problems facing households, and therefore reducing income instability is an important motive for income diversification associated with seasonality (Ellis 1998: 11). As Ellis adds, this requires income-earning opportunities with seasonal cycles which are not synchronized with the farm's own seasons. Migration to other agricultural zones is an example of this.

### 2.3.2 Risk and Risk Management Strategies

Rural households face considerable risks in their livelihood process and these can affect income in such a way as to endanger consumption. In the literature, risk is often cited as one important motive for livelihood diversification (Bryceson 1996; Jock R. Anderson 2003). In economic terms, risk is the subjective probability attached by individuals or by the household to the outcomes of the various income-generating activities in which they are engaged (Edward Anderson et al. 1977, in Ellis 1998: 12).[17] With reference to Crole-Rees (2002), the following types of risk can be defined:

- Agricultural production risk (e.g., climatic, water management, reduced soil and pasture fertility, pests, seasonal labour deficit);
- Market risks (e.g., input and output price risks, commercial risks);
- Financial risks (e.g., farm debts);
- Personal or human risks (e.g., death, diseases);
- Institutional risks (e.g., change in macroeconomic institutions, market liberalization).

Climatic risk is one of the major risks related to *agricultural production risk* along with changes in soil fertility, pests and the risk linked to production technique and timing (Day et al. 1992, in Crole-Rees 2002: 9). The importance of these risks is dependent on the management capacities and decisions of the household (Jock R. Anderson 2003). The irregular rain pattern in many tropical developing countries affects output and yield levels (Gill 1991). The absence of rain at key periods of the agricultural season frequently leads to crop failures. This risk may be reduced where irrigation is possible. Climatic risk should not be confused with seasonality, though they are interlinked. The core difference is that seasonality is an expected and definite feature (e.g., the expected yearly monsoon), while a risk is linked to the probability of unexpected outcomes (e.g., drought or floods during the monsoon season).

Rural households face *market risks*, since the prices of inputs, but especially of output goods, are usually unknown at the moment households have to make production decisions. Their effect will depend on the household's market orientation and on market efficiency. Poor or non-existent infrastructure (e.g., bad roads, no communication infrastructure) renders the market more isolated and its access more risky and time-consuming, and therefore more expensive. Furthermore, highly covariate yields[18] augment the local trade risk as they lead either to general surpluses or to deficits. This applies especially when high transport costs in addition to other transaction costs are combined with the low agricultural productivity inducing a low market integration (Crole-Rees 2002: 10).[19] Other market risks are the risks from lack of insurance and/or credit market opportunities (section 2.3.5).

*Financial risks* depend on the way the farm activities are financed, and thus on debt obligations. Depending on the debt obligations and conditions, households will have to allocate a greater or lesser portion of their returns to paying off debts. An important distinction between financial and business risks (including agricultural production, personal and market risks) is that the effect of the former is carried over for one or many years. This is not the case with business risks (ibid.: 11).

*Personal or human risks* involve human beings. Crises in the household, such as the household splitting through marriage, divorce or separation, illness, accident, sickness and so on, are not predictable and households are generally uninsured against them. Depending on the timing of these events, household income might suffer, in either the short or the long term. Such events hit small households, as well as those that have low labour availability and/or low substitutionality, especially hard. Anderson says of major threats such as the death or serious disability of one of the principal partners of the farm business: 'this is often an insurable risk, but it is seldom clear what value should be attached to the insured event when buying insurance' (Anderson 2003: 170).

Finally, *institutional risks* may be caused by unpredictable changes in the institutional setting, for example, changes in regulations (e.g., on land rights or use of pesticides), the privatization of particular services (e.g., extension) or macroeconomic changes. As was mentioned earlier, market liberalization may increase price variability and hence price risk. Risks can obviously differ between households and/or in time.

In his review of possible responses to risk in poor rural households of developing countries, Jock R. Anderson (2003) has found the following *individualistic, on-farm* strategies: (1) the collection of additional information which reduces uncertainty; (2) actions (or inactions) which considerably reduce or avoid risks, such as an effective farm system monitoring or the application of cautious safety standards; (3) the selection of less risky production technologies; and (4) the diversification of farming activities or investments into off-farm activities, especially for the poorest farmers. There are however other strategies which involve *sharing risks* with others, such as (1) informal risk-pooling, including sharecropping arrangements, (2) external farm financing, sometimes with informal credit arrangements, (3) formal or informal insurance arrangements and finally, (4) marketing arrangements such as price-pooling, forward contracting (known as 'contract farming') or hedging on a futures market (ibid.).

Finally, there is an important discussion relating to the question of whether poor rural households are risk-averse or not. Roumasset et al. (1979, in Ellis 1998: 12-13) state that income diversification as a risk strategy usually implies a trade-off between a higher total income involving greater probability of income failure and a lower total income involving smaller probability of income failure. In other words, households are risk-averse, and they are prepared to accept lower incomes for greater security. However, some authors have contested this. Barrett et al., for example, attribute this behaviour to the existence of *economies of scope* in production.[20] For them, diversification across crops is less likely to be attributable to risk management than to economies of scope due to soil and water management and to heterogeneous land quality (fertility, drainage, slope, etc.) (Barrett et al. 2001: 323). Ellis (1998: 13) mentions that one of the critical features of income diversification for risk reasons is achieving a portfolio with low covariate risk between its components. A characteristic of rural livelihoods in developing countries is that most of the income-earning opportunities open to poor households, i.e. own farm production and agricultural wage labour, exhibit high correlations between risks attached to alternative income streams. In other words, if there is a drought or a flood in a particular area, all income streams are adversely affected simultaneously. While on-farm diversity can take some advantage of differences in the risk-proneness of crops or crop mixes to adverse natural events, the protection this affords is only partial. As Ellis states, only diversification into non-farm incomes can result in low risk correlations between livelihood components (ibid.).

### 2.3.3 Vulnerability

The notion of vulnerability is a concept which is often mentioned in the literature on risk, seasonality and coping, and which is closely linked to these concepts (Ellis 2000b: 62).[21] Vulnerability arises from the combination of *exposure* to a threat with *sensitivity* to its adverse consequences (Devereux 2001: 508). A threat can be linked to an external risk factor such as drought, price fluctuations or a sudden disaster. *Sensitivity* refers to the magnitude of a system's response (or the coping capability) to an external threat, and is determined, for example, by assets, food stores, or support from kin or community. For example, if a small change in the price of rice rapidly causes widespread undernutrition amongst a specific human population, then the livelihood system is particularly sensitive to even this minor shock. The concept of vulnerability can be further refined by the concept of *resilience*. Resilience refers to the ability of an ecological or livelihood system to 'bounce back' from stress and shocks, and diversity is an important factor contributing to resilience in both natural and human systems. It follows, then, that the most robust livelihood system is one displaying high resilience and low sensitivity (Ellis 2000b: 62-3).

The most vulnerable households are those that are both highly prone to adverse external events and lack the assets or social support systems that could carry them through periods of adversity (2000b). As Devereux mentions, although poverty and vulnerability are not synonymous, the poor face greater exposure to livelihood threats (e.g., they live in marginal areas) and are more susceptible to shocks because their asset holdings (productive assets, savings, human capital and social capital) are generally lower (Devereux 2001: 509). This fact has also been highlighted in a study on South Asia (World Bank 2002a). As mentioned above, vulnerability is closely connected to risk. In fact, it is determined partly by risk factors which are generic to groups of people who are connected geographically or by shared risk characteristics (*exposure*), and partly by risk factors which are specific to individuals or individual households (*sensitivity*) (Devereux 2001: 509).

The most vulnerable rural poor are those with few productive resources who are usually landless and rely on selling their physical labour to earn a living. Poor nutrition and health, as well as lack of education and skills, undermine returns to labour. Their vulnerability is linked to the fact that the poorest are often the worst paid and most casualized segments of the labour and commodity markets, engaging in a multiplicity of intermittent, seasonal and/or poorly paid activities to survive. In addition, their lack of assets deprives the poor of an adequate buffer against a crisis, leaving them vulnerable to a crash of their poor standard of living (Kabeer 2002: 591). In South Asia, vulnerability also has a social dimension (ibid.: World Bank 2002a). Social inequalities enshrined in gender conventions and caste systems can lead to vulnerability. Gender conventions are not class-specific, leading to deprivations and vulnerabilities which do not necessarily depend on the level of a household's income (see e.g., the fact that female participation in the labour force is much lower than male participants). The caste system can confine those from lower castes to poorly paid, niche jobs. Age can equally be a factor of vulnerability: looking after the aged and the infirm is increasingly viewed as a burden among poorer households. The very young tend to be vulnerable: their welfare may be sacrificed in times of scarcity in order to protect economically active household members (Kabeer 2002: 591). Gender and age differentiation can therefore constitute so-called *intra-household vulnerabilities*, meaning that members of the same household can face different levels of vulnerability.

### 2.3.4 Risk Strategies, Coping Behaviour and Adaptation

How do the rural poor prepare for and respond to livelihood risk and vulnerability? The literature would suggest that this is in two main ways: *risk management strategies* and *coping behaviour*. Repetitive coping behaviour over time is referred to in the literature as *livelihood adaptation* (de Haan 2000: 348). This section details these concepts.

#### *Risk strategies and coping behaviour*

Ellis (1998: 13) suggests that it is important conceptually to separate *exante* risk strategies from *expost* coping behaviours. The former is a voluntary action and a planned response to threats which could potentially occur, while the second is an involuntary action, i.e. an unplanned reaction to an unexpected livelihood failure. Hence the term 'risk coping strategies' should be avoided because it mixes both concepts (Alderman and Paxson 1992: 2; DFID and Farrington 2004: 8). However, the literature on risk management strategies and coping behaviour does not generally follow a clear analytical distinction between *exante* strategies and *expost* behaviour.

Corbett (1988) has developed a simple three-stage model of the sequential phases, which he sees as characterizing coping behaviour. Each stage reflects an increasing level of desperation: insurance mechanisms at first, then the disposal of productive assets, and finally, destitution. This three-stage model can be refined by distinguishing between *exante* risk management strategies and *expost* coping behaviour (Table 2.2). Thus, insurance mechanisms can be both risk-management strategies undertaken *exante* to reduce the likelihood of failure of primary production, or coping behaviours if practised *expost*. Normally, at the second or third level of desperation, one can assume that only coping behaviours are practised. At this stage, the principal source of income generation has failed expectations and producers have literally 'to cope' until the next harvest (Davies 1996: 47-8).

The literature generally concurs with Corbett's outline. It recognizes that households facing food shortages are forced to trade off short-term

TABLE 2.2: SEQUENTIAL USE OF RISK MANAGEMENT STRATEGIES AND COPING BEHAVIOURS

| | *Risk management strategies (exante)* | *Coping behaviours (expost)* |
|---|---|---|
| *Stage one: insurance mechanisms* | Changes in cropping planting practices<br>Sale of small-stock<br>Reduction of current consumption levels<br>Collection of wild foods<br>Increased petty commodity production | Sale of small-stock<br>Reduction of current consumption levels<br>Collection of wild foods<br>Use of inter-household transfers and loans<br>Migration in search of employment<br>Sale of possessions (e.g.,jewellery) |
| *Stage two: disposal of productive assets* | | Sale of livestock (e.g., oxen)<br>Sale of agricultural tools<br>Sale or mortgaging of land<br>Credit from merchants and moneylenders<br>Reduction of current consumption levels |
| *Stage three: destitution* | | Distress migration |

*Source*: Adapted from Corbett (1988: 1107).

consumption needs against longer-term economic viability. The sequence of the strategic or coping activity is determined not only by the *effectiveness* of the strategy in terms of bridging the food gap, but also by the *cost* and *reversibility* of each action. Strategies or coping behaviours which have little long-term cost are adopted first (such as food rationing and withdrawing savings). Strategies or coping behaviours, with a higher long-term cost, which are difficult to reverse are adopted later (e.g., selling the family's plough). Finally, distress behaviours (such as migrating off the land) reflect economic destitution and a 'failure to cope' (Devereux 2001: 512). In other words, households seek first to protect the future income-generating capability, even if current consumption is compromized. Kabeer makes the link between reversibility and vulnerability, stating that 'households which rely on strategies at the 'irreversible' end of the spectrum are likely to be the most vulnerable or exposed to crisis for longer' (Kabeer 2002: 594). Thus, where possible, households will avoid irreversible strategies because they leave them less able to recover and more vulnerable to the next crisis.

Coping is seen as one determinant of diversification by Ellis (1998; 2000b). In fact, the sequence of coping behaviour as a response to crises can involve searches for new income sources in an early stage, and, at a later stage, the enforced asset sales can irrevocably alter the future livelihood patterns of the family (Ellis 1998: 14).

### *Adaptation*

Adaptation (or adaptive strategy) is a further concept used in the coping behaviour context. It gives a temporal (or repetitive) dimension to the coping concept. To recall Davies' (1996) definition, adaptive behaviours are coping activities which have become permanently incorporated into the normal cycle of activities.[22] As Devereux puts it, coping 'strategies' are thus responses to adverse events or shocks, while adaptive strategies (i.e. adaptation) are adjustments to adverse trends or processes (Devereux 2001: 512).

De Haan also discusses the link between coping and adaptation:

> The contextual impact of climatic change, world market and politics is growing stronger and shocks and stresses appear more frequently, so that it is becoming important to shape the coping strategies more permanently. Thus, temporary coping mechanisms develop into permanent 'adaptive strategies'. (de Haan 2000: 348)

However, he finishes by placing a question mark next to what the time scale of adaptation will be:

> . . . I am inclined to think that, at present, new coping and adaptive strategies will have to be developed as responses to new shocks and stresses, even before stability in livelihood as a result of a previous adaptation has come within reach. (ibid.)

Adaptation is also a determinant of diversification. In fact, one potential outcome of adaptation is diversification, which explicitly draws on a variety of dissimilar income sources (farm, non-farm, remittances, etc.) as its main characteristics. However, adaptation can also mean finding new ways of sustaining the existing income portfolio. Adaptation may result in the adoption of successively more vulnerable livelihood systems over time (Davies 1996). But, as Ellis states, the prime motive and consequence of successful diversification by the poor is to reduce vulnerability (Ellis 2000b).

### 2.3.5 Market Failures

The lack of markets for credit, labour or land is often mentioned as an important determinant of diversification in the literature (Ellis 1998; Barrett et al. 2001; Crole-Rees 2002).

#### *Credit and insurance markets*

The availability of financial capital to purchase inputs and capital equipment (e.g., water pumps) for agricultural production has long been regarded as one of the critical constraints limiting potential gains in productivity in small-farm agriculture. The severity of this constraint is thought to reside in the poor functioning of rural financial markets in developing countries (Ellis 2000a: 295). In rural Asia, private money-lending exists, but tends to be associated with personalized transactions in interlocked markets which can place the borrower in a permanent state of obligation to the lender (Colatei and Harriss-White 2001).

Credit market failures provide another reason for diversifying livelihoods (Reardon 1997), for example, by investing cash funds generated outside agriculture in order to purchase agricultural inputs or farm equipment. Ellis (2000a: 296) states that this investment strategy has the potential to overcome the absence of lending facilities in rural areas, avoid paying high rates of interest on such funds as may be available from public or private sector financial institutions and avoid placing the individual or the family in a subordinate social relationship with a private moneylender. The evidence that cash funds generated outside agriculture are reinvested in agriculture is, however, challenged by other studies. Bichsel et al. (2005: 36) found that remittances sent from abroad to rural areas in Mexico, Kyrgyzstan and south India are mostly spent for daily consumption purposes, health expenses, life-cycle events, and only to a small extent on agricultural production.

Though the lack of credit opportunities is often mentioned as being a determinant of diversification, missing financial markets can also be a *hindrance* to diversification, as they can impede diversification into activities or assets characterized by substantial barriers to entry. On the other hand, if non-farm or off-farm options can be accessed easily but credit markets are incomplete or missing, then non-farm earnings can be a crucial means of overcoming the working capital constraints to purchasing variable inputs needed for farming (Barrett et al. 2001: 321).

It is important when writing about credit markets to point out that the existence of a financial market does not mean per se that all rural smallholders have access to it. Access to credit and financial savings is very dependent on the availability of collateral. For less wealthy people in rural areas, limited access to credit can, for example, prevent the acquisition of livestock which is necessary to diversify out of crop agriculture; it can also mean that they cannot purchase other assets they need in order to diversify into remunerative non-farm activities in manufacturing or commerce. These entry barriers to credit markets tend to leave the poor with less diversified asset and income portfolios, thereby forcing them to face both lower returns and a higher variability in earnings (Barrett et al. 2001: 325).

Some authors have linked the discussion on the consequences of risk behaviour (as pointed out earlier) and the need for self-insurance with the subject of financial markets (Jalan and Ravallion 1999; Barrett et al. 2001; World Bank 2002a: 38-42). When financial markets are complete, economic theory suggests that individuals will consume only the permanent portion of their income and save (or spend) any occasional/ passing positive (or negative) earnings. Otherwise, if they are risk-averse, they will purchase insurance to have a guarantee against income risk. However, because financial markets are generally incomplete in the rural areas of less developed countries, individuals must act outside financial markets to reduce consumption variability as a result of real-income variability. Diversification is thus a primary means of reducing risk, and it can consequently be understood as a form of personal insurance (Barrett et al. 2001: 322).

### *Labour and land markets*

Rural wage labour is omnipresent in south India. This stems from high rural population densities, an endemic shortage of land, and significant inequalities in landownership (Ellis 2000a: 295). These factors lead to the existence of large numbers of landless or near landless families, for whom wage labour on other farms is the chief means of survival. The sections on seasonality and risk document the important role labour markets have in reducing cyclical and insecurity threats to rural livelihoods. But labour markets also offer non-farm opportunities (i.e. diversification opportunities) for income generation, which can be differentiated through education (e.g., for salaried jobs in business or education), skills (e.g., in trading, vehicle repair, brick-making), gender (female opportunities in trading or factories *vs.* male wage work in construction sector) and location. Ellis (ibid.) puts this in economic terms: 'when the marginal return to labour time in farming for any individual falls below the wage rate or the return to self-employment attainable for that person off the farm, then, ignoring intra-household distributional issues, the household as a unit is better off switching that individual into off-farm or non-farm activities.' However, economic considerations of labour allocation are modified by social rules of access, both within the family and in the community, and these rules may result in the social exclusion of individuals and households from particular income streams (Ellis 1998: 12). This will be discussed in the section on the role of the local institutional setting.

The allocation of labour depends to a great extent on the availability of land and functioning land markets. As Barrett et al. (2001: 321) explain, a smallholder household which has much labour at its disposal but relatively little land will, in the absence of well-functioning land markets, typically allocate some labour to its own farm, and hire some labour out for off-farm wage employment in agriculture: 'Individuals rationally allocate assets across activities to equalize marginal returns in the face of quasi-fixed complementary assets (e.g., land) or mobility barriers to expansion of existing (farm or non-farm) enterprises.' For the poorest, this usually means forced or desperation-led diversification. The declining size of landholdings is another important determinant for part-time farming: 'One reason that this [part-time farming] occurs is the continued sub-division of land at inheritance, resulting in declining farm size, and this is especially prevalent in areas of high agricultural potential' (Ellis and Biggs 2001: 445). Given the generally poor situation of land availability in Asia, one can expect a similar situation there.

## 2.4 ROLE OF THE LOCAL INSTITUTIONAL AND ORGANIZATIONAL SETTING IN LIVELIHOOD DIVERSIFICATION

Livelihood diversification is a process which is embedded in—and influenced by—the wider institutional and organizational context. In recent years, there has been a growing tendency for change in this institutional and organizational context: for example, market-led growth, trade liberalization, privatization, or a reduced role for the state in the economic development of many developing countries. Within this context, it has become important to understand better what links rural households have to this institutional and organizational setting and how these linkages affect their livelihood strategies (Ellis 1998). This section defines the concepts and gives the theoretical background.

### 2.4.1 Defining Institutions and Organizations

Institutions and organizations, and their role in development, form a much-studied topic in development research (Mummert 1999; Nuijten 1999; Müller-Böker 2001; Marsh 2003). Different development research disciplines (with different theoretical traditions) have adopted different analytical approaches to institutions and, as Nuijten suggests, 'a quick look at the many subjects which come together under the umbrella institution makes clear the need to be explicit about the way in which we use the concept' (Nuijten 1999: no pagination).

Uphoff makes a clear distinction between *institutions* and *organizations*. As he states, confusion often arises because they refer to different but overlapping sets of social phenomena (1993: 614). For Uphoff, *institutions*, whether organizations or not, are 'complexes of norms and behaviours that persist over time by serving collectively valued purposes', while *organizations* (whether institutions or not), are 'structures of recognized and accepted roles' (1993: 614).[23] Institutions, like organizations, have to be understood as a matter of degree. Marriage, for example, may become more or less of an 'institution' depending on the scope and intensity of its recognition and acceptance by people as a legitimate complex of norms and behaviours, one which people feel obliged to accommodate and comply with (Table 2.3). This is true of any institution, whether or not it has an organizational form. Thus, *institutionalization* is a process, and organizations can become more or less 'institutional' over time to the extent that they enjoy special status and legitimacy for having satisfied people's needs and for having met their normative expectations over time (ibid.). Uphoff's work tends therefore to stress the *normative* aspects of institutions while stressing the *structural* aspects of organizations.

Another definition of *institutions* and *organizations* is given by the *new institutional economists*. North (1990) defines institutions as 'humanly devised constraints that structure human interaction' and also calls them the 'rules of the game'. For North, institutions are made up of formal constraints such as rules, laws, constitutions and their enforcement characteristics. As Scott mentions, this emphasis on enforcement mechanisms derives from the nature of the customary objects studied by institutional economists: they 'are likely to focus attention on the behaviour of individuals and firms in markets and other competitive

TABLE 2.3: EXAMPLES OF INSTITUTIONS AND ORGANIZATIONS AS OVERLAPPING SETS

| *Institutions that are not organizations* | *Institutions that are organizations, and vice-versa* | *Organizations that are not institutions* |
|---|---|---|
| *Examples* | | |
| Money | The Central Bank | A local bank branch |
| Marriage | 'The Family' | A particular family |
| Technical assistance | World Bank | A consulting firm |

*Source*: Adapted from Uphoff (1993: 616).

situations, where contending interests are more common and, hence, explicit rules and referees more necessary to preserve order' (Scott 1995: 53). As such, North belongs to the groups of scholars who look at institutions in a *regulative* way, underscoring that institutions constrain and regularize behaviour (ibid.: 51).[24]

In its World Development Report 2003, the World Bank (2002b: 38) defines institutions as 'the rules, organizations, and social norms that facilitate coordination of human action'. Thus, organizations are a subcategory of institutions, independent of whether organizations are 'institutionalised' or not (following Uphoff's concept). This approach has been criticized by de Haan and Zoomers. They point out that authors such as Ellis have clearly distinguished and defined *social relations*, *institutions* and *organizations* instead of heaping them together; social relations are distinguished from institutions and the latter from organizations (de Haan and Zoomers 2005). For Ellis, social relations refer to the social positioning of individuals and households within society. This social positioning comprises such factors as gender, caste, class, age, ethinicity and religion. With reference to North (1990), Ellis defines institutions as the formal rules, conventions and informal codes of behaviour that comprise constraints on human interaction (Ellis 2000b; 38). Some examples of institutions are laws, land tenure arrangements (property rights) and markets.[25] The role of institutions is to reduce uncertainty by establishing a stable structure for human interactions (North 1990: 6). Organizations, as distinguished from institutions, are groups of individuals bound by the common purpose of achieving certain objectives (ibid.: 5). Examples of orgainizations are government agencies (the police force, ministry of agriculture, government veterinary service), administrative bodies (the local government), NGOs, associations (farmers' associations) and private companies (firms). With Uphoff in mind (see above), some of these organizations can also grow to become institutions as they get institutionalized over time; that is, they enjoy a special status and legitimacy through having satisfied people's needs and having met their normative expectations. Table 2.4 summarizes the definitions of social relation, institution and organization, including some examples.

The definitions contained in Table 2.4 will be used in this work. In addition, all institutions and organizations which exist in a specific context are called the *institutional and organizational setting*.

TABLE 2.4: DEFINITION OF SOCIAL RELATION, INSTITUTION AND ORGANIZATION

| *Social relation* | *Institution* | *Organization* |
|---|---|---|
| *Definition* | | |
| 'The social positioning of individuals and households within society' (North 1990) | 'The formal rules, conventions, and informal codes of behaviour, that comprise constraints on human interaction' (Ellis 2000b, following North 1990) | 'Groups of individuals bound by some common purpose to achieve objectives' (North 1990) |
| *Examples* | | |
| Gender<br>Caste<br>Class<br>Age<br>Ethnicity<br>Religion | Laws (e.g., criminal law)<br>Land tenure arrangement (e.g., property rights)<br>Markets | Government agencies (police force, ministries)<br>Administrative bodies (local government)<br>NGO's and associations<br>Private companies (firms) |

*Source*: Author's table based on Ellis (2000b) and North (1990).

### 2.4.2 Local Institutional and Organizational Context

When researching institutions and organizations it is essential to limit the space under study. For this reason, many scholars tend to analyse institutions and organizations present at the *local level* rather than all institutions at all levels (Marsh 2003). This section defines what is meant by 'local'. Uphoff (1993: 608) identified ten levels ranging from the 'international level' to the 'individual level'.[26] Of these, three are commonly regarded as being 'local', namely, the *locality*, the *community* and the *group* levels.[27] The *locality* is likely to be the area served by a rural market town. It includes a set of communities which have trading, intermarriage and other cooperative links with one another, where there is some possibility that people are personally acquainted and they usually have some experience of working together. It is the area rural people refer to when asked were they come from. The *community* or *village* level corresponds to what is most often thought of as 'local'. The *group* level is generally at a lower level, i.e. more elementary in social organizational terms, than a community, though groups can cross village boundaries. From a socio-economic perspective, the basic characteristic of what is 'local' is 'that most people within a locality, community or group have face-to-face relationships and are likely to have multi-stranded connections—as members of a common church, as buyers at the same market, as relatives through extended families' (Uphoff 1993: 609). These three levels provide a better basis for 'collective action' than found, for example, at district or national levels. Households and individuals, though they do exist at the local level, are not considered by Uphoff as being local levels since—as he states—'they do not present the same problematic issues of 'collective' action found with groups, communities and localities' (ibid.). In the present work, the term *local institutional and organizational setting* refers thus to the specific manifestations of institutions and organizations at the local level, meaning in the locality, community and group under study. Often institutions and organizations may cross the boundaries of this level (Appendini et al. 1999: no pagination).

For analytical clarity, institutions and organizations can be classified in different ways. North (1991, in Marsh 2003: 3) uses the categories of *formal* and *informal* institutions. Formal institutions stipulate rules such as constitutions, laws and property rights, while informal institutions are generally agreed-upon arrangements or rules of behaviour such as sanctions, taboos, customs, traditions and codes of conduct.[28] North's classification is used in the present study. Another mode of classification is to separate them into the categories of 'state', 'market' and 'civil institutions' (Uphoff 1993; World Bank 2002). Ultimately, this leads to the differentiation between the public sector, the private sector and a sector in between called by some authors the 'third sector' (Uphoff 1993: 609; Sachs 2000: 136-42). This third sector can be designated 'the voluntary sector, the membership sector, the self-help sector, the participatory sector, or the collective action sector' (Uphoff 1993: 609).

### 2.4.3 Linking the Local Institutional and Organizational Setting with Diversification Processes

How does the local institutional and organizational context link with processes of livelihood diversification? To conceptualize these linkages, the concepts of transaction costs, exclusion, inclusion and access, and bonding and bridging institutions are useful.

### *Transaction costs*

The concept of *transaction cost* derives from the work of the new institutional economists who have criticized both neoclassical and neo-Marxists economic theories for their lack to take into consideration the costliness of market exchange (Ensminger 1992). Arrow (1969: 48, in Williamson 1985) defines transaction costs as the 'costs of running the economic system'. These costs have to be distinguished from production costs, which is the category of costs with which neoclassical economists have been preoccupied. Williamson (1985: 19) states that 'transaction costs are the economic equivalent of friction in physical systems'. But while physicists are quickly reminded by their laboratory instruments and the world around them that friction was pervasive and often needed to be taken into account, economists do not have a corresponding appreciation for the costs of running the economic system (Williamson 1985). As an example, the exchange of agricultural and livestock products can be affected by the following transaction costs: spoilage, quality differences depending on processing, lumpiness of initial investments, lags in production, seasonal variability, search costs, screening trade partners, bargaining, monitoring, or contract enforcement (Delgado and Siamwalla 1997). Today, with the rise of modern food systems, a new set of transaction costs has arisen because of the standards required in terms of quality, size and delivery (Pingali et al. 2005: 64). And these new transaction costs do not favour smallholder farmers (Kydd and Dorward 2001: 471).

Within communities (or within the local level defined above), transaction costs tend to be low because of the predominance of traditional forms of exchange based on indigenous institutions such as reciprocity and redistribution, and because of personalized forms of market exchange in which individuals are known (if not related) to one another and engage in repeated dealings. However, potential gains from small-scale exchange within communities are severely limited because the trading partners are likely to face the same environmental conditions and often produce the same products. In order to capture higher comparative advantage, therefore, broader trade and specialization are necessary, yet this implies an increase in transaction costs (Véron 1999: 49). When the additional transaction costs of the exchange process outweigh the gains from exchange, societies do not benefit from trade and specialization (Ensminger 1992).

Transaction costs can be dependent both on the *size* and on the *location* of a farm. In some countries, small and large farm households in rural areas often do not have access to the same technology, information, asset base, input supplies and market outlets. And, rural households with different assets' bases are likely to face different levels of transaction costs: poorer households may have more difficulty diversifying into new activities than more wealthy ones. The same can be true for households situated in different locations. In remote areas where physical access to markets is costly due to heavy transaction costs, households diversify production patterns, partly to satisfy their own demand for a certain diversity in consumption (Omamo 1998). Thus, under these different conditions, different farm households are likely to be subject to significantly different levels of transaction costs in order to produce and sell the same output (Delgado and Siamwalla 1997: 4). Furthermore, high-value farm products (e.g., animal protein or horticulture products), which often figure in agricultural diversification, are characterized by a high ratio of transaction costs to their final value.

Delgado and Siamwalla (ibid.) write that 'lowering and reducing disparities of transaction costs across rural households is . . . central . . . to promoting farm-level adjustment of agricultural output mixes to major changes in relative prices'. Some important institutions to reduce transaction costs are vertical integration of production and marketing, contract farming, and various forms of producer cooperatives or village self-help groups. Jones Killick also sees the level of transaction costs (linked to institutional and Organization development) as an important determinants of market access (2001: 171). Finally, the World Development Report 2003 puts emphasis on the importance of coordinating institutions for lowering transaction costs (World Bank 2002: 39).

### *Exclusion, inclusion and access*

There are two types of approachs to *social exclusion* (Gore 1994, in de Haan 2000: 351). The first approach interprets social exclusion as a mechanism by which elites exclude others from access to resources with the objective of maximizing their own returns. This view sees social exclusion as a process by which groups try to monopolize specific, mostly economic, opportunities to their own advantage. They often use certain social or physical characteristics such as race, gender, language, ethnicity, origin or religion to legitimize depriving others of opportunities. As an example, a specific rural woman will never have her own plot if others are able to deny her land, and she will never grow vegetables if others can deny her irrigation water. The second approach to social exclusion comes from marginalization studies. While, in the first view, exclusion literally means 'being excluded', in marginalization studies it rather means 'being left behind' after a failed attempt at inclusion. As de Haan and Zoomers (2005: 34) conclude, 'livelihood activities are not neutral, but engender processes of inclusion and exclusion'. These processes are strongly shaped by institutions and organizations; they 'are part and parcel of any Organization and institution' (Nuijten 1999: no pagination). Understanding the relationship between institutions and access to resources and other income-generating activities needs a study of the institutional dynamics and institutional processes of inclusion and exclusion. Results from an FAO study on 'local institutions supporting sustainable livelihoods' (Marsh 2003) show that local economic institutions are typically more inclusive than formal, extra-local institutions and projects which require participants, members and/or managers to have a minimum level of education and wealth (e.g., rural banks, agricultural extensions, and even certain anti-poverty schemes). As Marsh writes, 'non-local interventions may simply bypass the poor and other marginalized groups because their needs and capacities are invisible to outsiders' (2003: 26). The study, however, does not furnish more details of the institutional process that leads to exclusion or inclusion.

In the economic literature, a concept similar to exclusion is the one of *entry barriers*. This concept is used to show the difficulties of farm households to, for example, enter into certain economic activities due to specific economic, institutional or political constraints (Reardon et al. 1998; Abdulai and Crole-Rees 2001; Barrett et al. 2001; Davis et al. 2002; Davis 2004).

*Access* is another important concept close to those of exclusion/inclusion and entry barriers. According to Merriam-Webster (1993, in

Ribot 1998: 310), *access* is the 'freedom or ability to obtain or make use of'. One crucial difference relating to access is that between the terms *right* and *ability*. The term 'right' implies an acknowledged claim that society supports (whether through law, custom or convention). The term 'ability', however, is broader than 'right', 'resting solely on demonstration without the need for any socially articulated approval' (Ribot 1998: 310). When one refers to access, the question being posed is, access to what? In this work, access is used in relation to institutions and organizations and their role in hindering or supporting diversification processes. Therefore, access is understood as the ability and/or right to obtain new (and diversified) income-generating activities. This ability and/or right is usually given through institutions and organizations which have the potential to support the process.

Access is based on a broad set of factors including the control of access through *power* (Nuijten 1999). The important role of power relations in livelihood outcomes has also been shown by de Haan and Zoomers (2005: 36-7). Power can be an important factor in allowing households to access new income opportunities. For example, if a farmer wants to sell his vegetables in the profitable urban market, he needs access to transparent market prices or to a trustworthy trader to sell to. Thus, access also depends on the performance of social relations, and these are sometimes far from harmonious (de Haan and Zoomers 2005: 34).

### *Bridging and bonding institutions*

The concepts of *bridging* and *bonding* are often heard in relation to institutions. These originally derive from a perspective on social capital which attempts to account for both its upside and its downside; they stress the importance of vertical as well as horizontal associations between people and of relations within and among such organizational entities as community groups and firms (Woolcock and Narayan 2000: 230). *Bonding* institutions are institutions that have the ability of linking people who are similar in terms of their demographic characteristics, such as family members, neighbours, close colleagues or farm colleagues. Thus, they build on the strong intra-community ties that give families and communities their sense of identity and common purpose. Bonding local institutions—based on traditional norms of solidarity and reciprocity—

are key elements in household livelihood strategies and community cohesion (Marsh 2003: 25). Together, these institutions (e.g., traditional authorities, common property rights, asset-sharing arrangements, mutual assistance, informal saving and credit, etc.) ensure a minimum level of food security and a safety net for the most vulnerable groups. They govern access to communal natural resources which provide sustenance (especially for the poor and in times of crisis), and they organize and finance community public works and civic-religious-cultural festivals and traditions. Their relative strengths and weaknesses are often associated with greater or lesser levels of social capital and intra-community conflict. Interestingly, evidence does not show a strong link between participation in these types of bonding institutions and higher levels of income and consumption or of social mobility (ibid.).

*Bridging institutions* connect people who do not normally share common characteristics. They are heterogeneous institutions bringing together members of different social and economic positions and influence, often with outside assistance. Michael Woolcock and Deepa Narayan stress that, without such intercommunity ties that cross social divides (based on religion, class, ethnicity, gender and socio-economic status), strong horizontal (bonding) ties can become 'a basis for the pursuit of narrow sectarian interests' (Woolcock nad Narayan 2000: 230). Bridging institutions can be an effective means for the poor to achieve some social mobility (Marsh 2003: 25). He cites examples of dairy cooperatives (with related cattle-sharing arrangements), migrant associations, watershed-user committees, political parties, churches and local governance institutions. These institutions afford their poorer members an opportunity to gain access to information and technologies, as well as NGO and government initiatives, which they may not otherwise learn about (ibid.). However, Marsh also states that power inequities within these bridging institutions can often prevent the full participation of the poorer members.

Looking at economic development Woolcock and Narayan (2000) argue that poor entrepreneurs, as an example, initially depend on their immediate neighbours and friends for credit, insurance and support (bonding institutions), while at a later stage, they require access to more extensive product and factor markets as their businesses expand (bridging institutions). Thus, 'as community members' welfare changes over time, so too does the optimal calculus of costs and benefits associated with particular combinations of bonds and bridges' (Woolcock and Narayan 2000: 231).

## 2.5 EFFECTS AND TYPOLOGIES OF LIVELIHOOD DIVERSIFICATION

This section first reviews the literature on effects and outcomes of rural livelihood diversification before presenting different typologies of diversification. Finally, the process approach to diversification—showing also the importance of structural components—will be highlighted. The focus of this section, as in the previous ones, is on results at the household level and thus excludes, the literature dealing with concepts and approaches at other levels, such as the rural sector level.

### 2.5.1 Positive Effects of Diversification

The literature on diversification describes the different effects and outcomes of the diversification process. However, the relationships between diverse income portfolios, levels of income, and income distribution are quite complicated, and, as Ellis has advanced, they are more prone to unhelpful generalization than illuminating specific contexts (Ellis 1998: 17).

The more straightforward effects of diversification on incomes follow from the determinants of diversification discussed earlier. Following Ellis (1998: 17), diversification might be expected to (1) reduce the overall risk of income failure by diluting the impact of failure in any single income source; (2) reduce *intra-year* income variability by diluting the effect of seasonality on farm-based income streams; and (3) reduce *inter-year* income variability resulting from instability in agricultural production and markets. These effects can be potentially beneficial for poor and rich households alike. However, the desperation-led diversification of the very poor can sometimes result from the accumulation-led diversification of the rich; for example, land consolidation by richer farmers in order to use capital-intensive agricultural methods can lead to a lack of land for the poorest.

Empirical evidence suggests that the effects of diversification play a significant *food security role* at the household level by ensuring food consumption in the deficit season (Reardon et al. 1992). However, this effect depends on the type of income source: *off-farm agricultural wage* income is likely to be correlated with the same seasonal and risk factors as own-farm production, and is therefore less able to play a role in evening out unstable income streams. *Local non-agricultural wage employment* and some types of *non-agricultural self-employment* (e.g., marketing, food processing, input supply) will also exhibit seasonal patterns similar to those experienced in agriculture. On the other hand, rural small-scale industries may be able to play a counter-cyclical role, and remittances from urban areas or abroad are likely to be independent of seasonal cycles in agriculture (different authors in Ellis 1998: 18).

In his review of the literature, Reardon found a strong positive relation between the *share* of non-farm income and total household income, and therefore an even more pronounced relationship between the *level* of non-farm income and total income (Reardon 1997). A review by Barrett et al. (2001) of the effects of diversification on income and wealth confirms these findings and shows that non-farm activity is typically positively correlated with income and wealth (in the form of land and livestock). This does therefore seem to offer a way out of poverty for the rural poor if they can seize non-farm opportunities. However, this key finding is 'a double-edged sword', since it also suggests that those who begin poor in land and financial capital have to overcome entry barriers and steep investment requirements to be able to participate in non-farm activities capable of lifting them out of poverty.

Besides having an impact on income and wealth, the literature also notes other positive effects of diversification. Some authors argue that non-farm income sources injected into the farm household can be an agent for positive change in their agricultural production. For example, remittances from migrant family members are used for agricultural investment and they can thus be interpreted as relieving a credit constraint caused by credit market failures. Non-farm income can also be viewed as a substitute for insurance, enabling the farm household to carry out risky innovations (different authors in Ellis 1998: 21-2). Another advantage of off-farm earnings is that if land is unacceptable as collateral for a loan, a regular non-farm cash income might be enough to enable one to borrow. Finally, non-farm earnings may be essential in allowing farmers to maintain a viable farm when they have a deep attachment to agriculture as a way of life and are willing to pay the price of farming in the form of foregone profits (Barrett et al. 2001: 321-2).

### 2.5.2 Adverse Effects and Critiques of Diversification

As has been mentioned, in the foregoing, Reardon (1997) and Barrett et al. (2001) suggest that it can be very difficult for poor farmers lacking financial capital and land resources to enter into diversification. In fact, those with the least agricultural assets and income are typically also the least able to compensate for this deficiency through non-farm earnings as they cannot meet the investments needed to enter into remunerative non-farm activities. Thus, they ask the critical question whether it is only higher incomes that open the door to attractive non-farm opportunities, and suggest that 'there seem to exist substantial entry or mobility barriers to high return niches within the rural non-farm economy' (Barrett et al. 2001: 324).

Entry barriers manifest themselves in the form of labour-market dualism: 'the skilled and educated are self-employed or can secure stable long-term employment at relatively high salaries, while the unskilled and uneducated depend disproportionately on more erratic, lower paying casual wage labour, especially in the farm sector' (ibid.). Thus labour-market dualism is directly linked to the educational achievements of household members: this proves to be the most important determinant of non-farm earnings, especially in more remunerative salaried and skilled employment. On the other hand, skills and educational achievement represent substantial entry barriers to high-paying non-farm employment or self-employment (325). Dercon and Krishnan confirm the importance of human capital: 'better off families are able to diversify in more favourable labour-markets than the poor. Important reasons for this effect are the lack of assets of the poor and their exclusion from the more highly remunerated labour markets due to skill and educational constraints.' (1996, in Ellis 1998: 19)

Along with the problem of entry barriers, several other negative effects of diversification are cited in the literature. According to a recent quantitative study in India (Anderson and Deshingkar 2005), income diversification has an adverse effect on households' earned income. The results show that individuals based in households with more diversified incomes have lower overall incomes than individuals based in more specialized households. Though the difference is quite small, they hypothesize that it is due to the costs involved in the diversification process. Another reason for unsuccessful diversification is that adopting many livelihood activities can reduce the economies of scale that are generated through specializing in a particular activity (Ashley et al. 2003). Thus, there is a potential trade-off between diversity and stability of income; in fact, diversified household livelihoods can be susceptible to a wider range of uncertainties and shocks, even though they may not happen at the same time. This is, however, in contradiction with the positive effects on income and wealth mentioned above.

Finally, an important criticism levelled at household diversification is the adverse effects it can have on farm investment and output. Pottier mentions the fact that in some cases farm profits are invested in non-farm enterprises rather than the other way round (Pottier 1983, in Ellis 1998: 21-2). A further constraint is that non-farm diversification takes able-bodied and skilled labour out of farm production. As Ellis says, 'it is typically the younger, more innovative, better educated members of farm families that leave the farm to engage in rural non-farm activities or to undertake distance migration' (Ellis 1998: 21). Studies on remittances from international labour migration also show that they are mostly utilized by the household for consumption purposes rather than productive investment into agriculture (Bichsel et al. 2005: 37-9).

### 2.5.3 Diversification Typologies

Analysis of diversification call for a classification of the different ways in which households can potentially diversify. Such a classification can be found in many studies (Zoomers 2001; Davis and Bezemer 2004; Meert et al. 2005).

#### *Demand-pull and distress-push*

One approach that is sensitive to the different potentialities of rural diversity is the distinction between *demand-pull* and *distress-push* diversification (Davis and Bezemer 2004: 9).[29] *Distress-push* diversification typically occurs in an environment of risk, market imperfections and agricultural unemployment, and is typically triggered by an economic situation which sets the household on a downward income trajectory. It implies that engaging in economic activities which are less productive than agricultural production could happen on a full-employment basis, and is motivated by the need to avoid a further decline in income. Thus, *distress-push* diversification attracts households that are less well endowed or have lower incomes. Income is often gained through wage employment. *Demand-pull* diversification, on the other hand, is characterized as a response to evolving market or technological opportunities which offer the potential for increasing labour productivity and household incomes. *Demand-pull* attracts richer households, which can afford to invest in higher-return activities, which are often non-agricultural and a part of their own independent enterprise (Davis and Bezemer 2004: 9).

The use of the concepts of 'pull' and 'push' has been criticized by Ellis. He writes that while it is occasionally useful to dichotomize cause and effect in this way, 'the ease of so doing should not be confused with the accuracy of the description thus achieved' (Ellis 1998: 17). In fact, experience shows that individuals and households are influenced by a multiplicity of factors determining their livelihood changes. While on occasion a single factor may predominate over all others for an entire community or location, more often than not it is a cumulative combination of factors which will represent variable pressures and opportunities for different individuals and households within the community (ibid.).

### *Income-driven and activity-driven*

The literature on non-farm diversification contains another typology which focuses solely on the diversification into the non-farm sector (Davis and Bezemer 2004: 10). One can identify two main components to the process of diversification: income and activity. The *income-driven*, non-farm diversification hypothesis assumes that diversifiers are profit-maximizers, while the second *activity-driven*, non-farm diversification highlights the different comparative advantages of household members as underlying incentives for non-farm diversification (Ellis 1993 in Davis and Bezemer 2004: 10). *Income-driven* diversification coincides with a period of capital accumulation, while the second type, activity-driven diversification, often occurs at a later point when the aforementioned capital accumulation has already taken place. Davis and Bezemer (2004: 10) see it as a mixed and dynamic process, with *income-driven* and *activity-driven* diversification often overlapping or occurring at the same time. They add that these are two stages in a process which is not necessarily sequential, but cyclical. Firstly, the income-dominant phase is associated with the aim of covering a household's basic needs. This phase will be dominant as long as meeting basic needs is the household's main priority, as reflected by low levels of income. When incomes are securely above a particular threshold, a certain amount of capital may have been accumulated (a consequence of *income-driven* diversification). This enables the development of *activity-driven* diversification, allowing household members to select particular activities according to their comparative advantages, freed from the necessity of catering for basic needs by whatever means they can (Davis and Bezemer 2004: 10-11).

### 2.5.4 Process Approach to Diversification

The positive and adverse effects of diversification described thus far reflect relatively general and conclusive effects when they are the result of a well-thought-out diversification 'strategy'. This following highlights the role of structural components in the diversification process, alongside the more 'strategic' behavioural ones. It also presents the concept of the 'pathway' as part of the process approach to diversification.

Many authors have recently questioned whether households actually pursue strategies. To quote Bryceson:

> Rural income diversification and the sustainable livelihoods literature generally assumes that income diversification is a household-based livelihood strategy premised on members' complementary activities. But are rural households' activity portfolios a result of calculation and planning and do they reflect household cooperation and authority? . . . assumptions regarding the existence of collective or agreed household decision-making aimed at optimizing the allocation of household members' labour between activities must be treated with caution. (Bryceson 1999: 183-4)

As de Haan and Zoomers (2005: 38) put it in their analysis, the word 'strategy' arises out of the focus of contemporary livelihood studies on the active involvement of people in responding to and enforcing change. Rather than being seen as victims, they play an active role in managing their livelihoods. From this perspective the idea of capturing behaviour in terms of strategies is understandable. However, some trends have emerged that raise questions about this 'strategic' view of the behaviour of households. Household studies have, for example, shown that intra-household differences do exist and that the household cannot be considered a homogeneous unit of corresponding interests acting strategically (Wilson 2004; de Haan and Zoomers 2005).[30] Household behaviour is therefore not always deliberate or conscious, although 'people constantly weigh different objectives, opportunities and limitations in response to external and internal circumstances that change over time' (de Haan and Zoomers 2005: 38-9). To cite Schmink: 'the concept of . . . strategy can lose its meaning to the extent that it becomes a mere functionalist label applied ex post to whatever behaviour is found' (1984, in de Haan and Zoomers 2005: 39).

Some authors have acknowledged that behaviours—whether they are intentional or unintentional—have to be put in their specific contexts (de Haan and Zoomers 2005; Krishna 2005). Zoomers (1999, in de Haan and Zoomers 2005) mentions, for example, that geographical settings (whether rural or urban), different agro-ecological zones in mountainous regions and distance to markets can influence the set of opportunities and outcomes. The success of farmers may be less related to strategic actions than to location-related factors. Seasonality also plays an important role, and the household life cycle can be a major determining factor. The importance of structural factors is also highlighted in a study by Anirudh Krishna, when he states that what matters for success in escaping poverty varies considerably from one location to another, even within the same state and region (2005: 13).

In her studies about rural livelihoods in the Andes, Zoomers defines four categories of 'strategies' which—as she acknowledges—also depend on structural components: accumulation, consolidation/income maximization, compensation and survival, and security and risk-reduction (Table 2.5). *Accumulation* strategies involve establishing a minimum resource base to obtain the minimum manoeuvring space necessary to prepare for future expansion and upward social mobility. They are applied by recently married couples and families with young children, and include strategies such as migration, land acquisition and labour recruitment. *Consolidation/income maximization* strategies are applied by wealthier households and follow a period of upward mobility. The aim is to stabilize the household's well-being. These households—with many elderly family members—have surplus assets to invest, and their major strategy is land improvement. *Compensation and survival* strategies prevail among farmers who have to deal with a sudden shock (e.g., crop failure) or with a structural shortage of land or labour force (e.g., incomplete households).

TABLE 2.5: LIVELIHOOD PROCESSES IN RESPONSE TO INCREASING LEVELS OF PROSPERITY AND FAMILY DEVELOPMENT CYCLE

| *Dominant strategy and goal* | *Sub-strategies* | *Type of farmer/stage in family development cycle* |
|---|---|---|
| *Accumulation* | | |
| To establish a minimum resource base and prepare for future expansion | Temporary migration<br>Land acquisition<br>Labour recruitment | Newly-weds and young families with small children, who are relatively poor but seek to accumulate resources for improvement |
| *Consolidation/income maximization* | | |
| Consolidation and quality improvements; achieving qualitative improvement of production means | Land improvement/ agricultural intensification<br>Education<br>Consumption<br>Housing/residence | Consolidation households (children have grown up) that have achieved a period of upward mobility (fairly wealthy) |
| *Compensation and survival* | | |
| To break out of the slump or at least to remain afloat in their current situation (survival) | Migration<br>Saving and economizing<br>Selling Exchange, borrowing and barter<br>External support | Disintegrating households experiencing downward social mobility and/or households experiencing a temporary crisis |
| *Security and risk-reduction* | | |
| (Often combined with other strategies above) | Diversification<br>Sharecropping<br>Stockpiling | All types, but especially poorer households |

*Source*: Adaptation based on Zoomers (2001: 251).

These households experience downward social mobility and try to find a way up (again) through migration, saving, selling capital assets, borrowing and bartering, and reliance on family social security. Finally, *security and risk-reduction* strategies are common in areas that are less secure for ecological reasons (e.g., drought-prone regions). These strategies include diversification through multi-cropping and multitasking, and exploring non-agricultural opportunities (de Haan and Zoomers 2005: 39-40).

Zoomers stresses that these categories should be looked at in a flexible way since a livelihood strategy is a 'moving target'. Thus, any given strategy should be conceived as a stage rather than as a structural category (Zoomers 1999, in de Haan and Zoomers 2005). She adds that instead of classifying farmers on the basis of what they own, it would better to characterize them in terms of their objectives and priorities, which means also by elements of their personality. Finally, families have different starting points, and processes of upward and downward social mobility are less unilinear than many assume. While the poor are generally expected to experience further deterioration in their situation, the rich are presumed to be in a favourable position for improvement. However, Zoomers found substantial numbers of people who used to be poor but are now rich, and people who used to be rich but are now poor (Zoomers 1999 in de Haan and Zoomers 2005: 39). A study on poverty in India confirms this: in one particular region, 14 per cent of the households escaped from poverty during the last 25 years, but 12 per cent fell into poverty during the same time (Krishna 2005).

In addition to the structural components mentioned above, de Haan and Zoomers (2005: 40) mention the importance of socio-cultural components for the way they introduce aspects of power and access into the livelihood research. In doing so, they also introduce the concept of pathways. *Pathways* are defined as patterns of livelihood activities which arise from a process of coordination amongst actors. 'This coordination emerges from individual strategic behaviour embedded both in a historical repertoire and in social differentiation, including power relations and institutional processes, both of which pre-structure subsequent decision-making' (de Haan and Zoomers 2005: 43).[31] 'Pathways show that people do shape their own livelihoods, but not necessarily under conditions of their own choosing' (ibid.). As Ramisch et al. (2002: 183) state, 'livelihoods emerge out of past actions and decisions are made within specific historical and agro-ecological conditions, and are constantly shaped by institutions and social arrangements'.

## 2.6 ANALYTICAL AND CONCEPTUAL FRAMEWORK

As this chapter shows, the study of diversification in rural livelihoods has grown in importance since the 1980s, and is directly connected to the theoretical debate about the role of the rural non-farm economy in rural development. Livelihood diversification is defined here as 'the process by which rural families construct a diverse portfolio of activities and social support capabilities in their struggle for survival and in order to improve their standards of living' (Ellis 1998).

The present work is based on the analysis of data from a case study of the income-generating activities of rubber-holding households in Thalanadu, Kerala, in south India. This section summarizes the analytical and conceptual framework that is used to analyse the empirical data (Fig. 2.1).

The main study unit used are the households of rubber holdings within the study area (section 5.2). The object under study is the portfolio of income-generating activities of each rubber-holding household during the financial year 2003-4. To understand whether households—and which of them—have diverse livelihood strategies, a first analysis concentrated on quantifying and categorizing the different income sources. This was done using the categorization suggested by Barrett et al. that separates income and activities in agricultural and non-agricultural (sector), on-farm or off-farm (space) and wage-employment or self-employment (function). Taking into consideration (1) the yearly gross

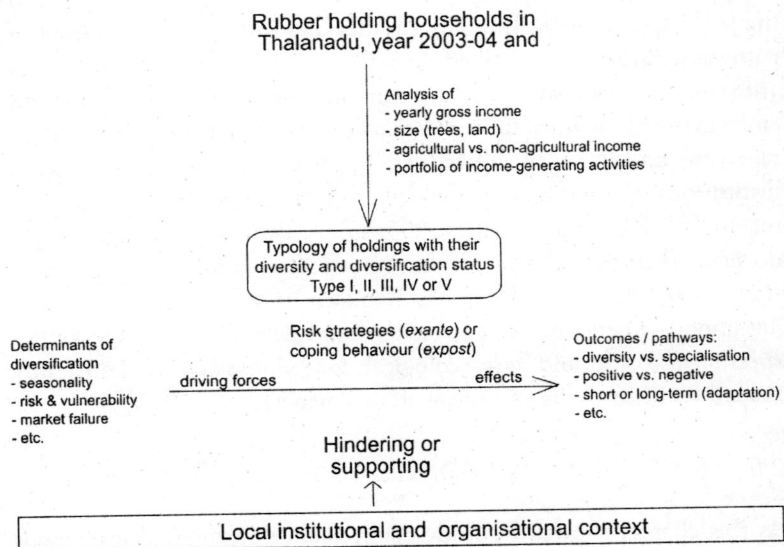


*Source*: Author's figure.

Fig. 2.1: Analytical and conceptual framework.

income of the holding, (2) the size of the holding (total land ownership), and (3) the relative importance of agricultural *vs.* non-agricultural income, a typology was developed which separated the rubber holdings into five specific types. This typology brings together holdings which do not have similar income and activity patterns; this allowed an assessment of the importance and the role of diversified income and activity in opposition to specialized ones.

It is crucial when analysing diversification as a temporal process to understand which driving forces (or determinants) are at the root of the diversification process. An inter-temporal analysis also has to be made; in this work, it is based on qualitative data. The data from the study area makes it possible to show the importance of seasonality, risk and the vulnerability context, as well as the role of the failure of markets (for credit, insurance, labour and land). The analysis of each single determinant brings with it the danger of a rather one-dimensional view of the many forces shaping rural livelihoods. Ellis points to the fact that these determinants may be conceptually distinct but are rarely separable in practice (Ellis 1998: 16).

But, how do households prepare for and respond to these determinants? The literature differentiates between the role of *risk strategies* and the one of *coping behaviour*. The former is a voluntary and planned action to threats which could potentially occur, and thus *exante*. The latter is an unplanned reaction to an unexpected livelihood failure due to internal (e.g., disease of a household member) or external factors (e.g., heavy drought), and is thus an *expost* reaction. *Exante* risk strategies and *expost* coping behaviour are, however, themselves determinants of diversification since they favour the construction of a diverse portfolio of activities and social support system, for example, via the changes in cropping pattern, or the reduction of current consumption levels. If, over a longer time, households regularly show the same patterns of coping behaviour, the concept of *adaptation* is used.

Livelihood strategies are not a fixed category with a final objective, but rather they are in a continual state of flux, which depends on the point of the households' starting point, the family's development cycle and other factors. Thus, it is difficult to speak of livelihood strategies in terms of *outcomes*, including diversification. This work limits itself to a broad analysis of some positive and negative effects on households which are at a diversified stage. These households can be classified using

Zoomers' categorization. Finally, the concept of *pathways* is useful in analysing that (diversified) livelihoods are connected to a historical and sociocultural context, as well as being shaped by institutions and social arrangements.

As is shown in Fig. 2.1, this analysis would be incomplete without looking at the role of institutions and organizations found in the particular context. As has been shown by many studies, livelihood diversification is actually a process which is embedded in and influenced by the wider institutional and organizational context. This context is itself not fixed, but experiencing constant change and adaptation to new trends (e.g., market liberalization, privatization). To make the analysis comprehensible, it focuses on institutions and organizations acting or present within the local context (as defined by Uphoff). So how does this *local institutional and organizational setting* link up with processes of diversification? A first way in which this question is addressed by the study is through the concept of *transaction costs* (with reference to Williamson and Ensminger). Transaction costs—which are dependent both on the size and the location of a holding—can be a hindrance for households who wish to change their livelihood portfolio from one set of activities to another. Another way of studying the impact of the institutional and organizational setting is by applying the concepts of *exclusion* and *inclusion*, and *entry barriers* or *access*. These concepts highlight the fact that institutions and organizations can hinder households or families, either voluntarily or involuntarily, in their access to certain activities or incomes (when, for example, a household is 'being left behind' and thus marginalized). Entry barriers and access are concepts similar to each other. Finally, the concepts of *bridging* and *bonding* institutions (following Woolcock) are also applied. The local institutional and organizational setting can hinder or support diversification processes in two ways: by influencing the risk strategies and coping behaviour which potentially leads to diversification, or by influencing the effects of diversification.

## NOTES

1. Ellis and Biggs note that these strands are all interlinked, even if they emerge from different places. In fact, 'while advocates of grass-roots approaches to development may like to think that they have nothing in common with World Bank market liberalizers, nevertheless the spaces in which grass-roots action flourished from the mid-1980s onwards were created in some measure by the

backing off by big government from heavy-handed involvement in the rural economy (Ellis and Biggs 2001: 443).

2. For transaction costs, see section 2.4.3.
3. Initially a lower dependency ratio due to falling birth rates, but then a higher one, as migration and AIDS remove young adults (Ashley and Maxwell 2001: 400).
4. Datt and Ravallion found, for example, that in India rural growth reduces poverty both in rural and urban areas, but on the contrary, urban growth does not alleviate poverty in rural areas (1996, in Ashley and Maxwell 2001: 403).
5. Devereux also goes into this direction by stating that 'where regular cash transfers are provided to the poor, the impacts are magnified by income multipliers, investment in agriculture and family enterprises, informal redistribution to relatives and friends, and stimulation of local trade' (Devereux 2001: 515). Against the powerful arguments of the high costs involved in such a 'solution', Devereux counters that redistributive transfer programmes are alleged to be fiscally unaffordable and unsustainable—though, in reality, all public spending allocations are policy choices (516).
6. For studies on the rural non-farm economy, see Davis and Bezemer (2004) for conceptual issues; Lanjouw and Lanjouw (2001) and R. Davis (2004) for global studies; Lanjouw and Shariff (2004) for a review specific to India. On farm/non-farm linkages, see Davis et al. (2002).
7. The first paragraphs of this section are based on de Haan and Zoomers (2005: 28-30).
8. The household as a unit also became popular in a more practical sense; it was considered a convenient unit for the collection of empirical data (de Haan and Zoomers 2005: 28).
9. There are also insightful approaches from the 'North', such as the study by Meert et al. on farm household strategies and diversification on marginal farms in Belgium (Meert et al. 2005).
10. This section is mainly based on Ellis (1998: 4-5).
11. For Ellis, income refers to the cash earnings of the household plus payments in kind which can be valued at market prices. The cash earnings component of income includes items like crop or livestock sales, wages, rents, and remittances. The in-kind component of income refers to consumption of one's own farm produce, payments in kind (e.g., in food), and transfers or exchanges of consumption items which occur between households in rural communities (Ellis 1998: 4).
12. In Indian literature, the term *occupational diversification* is often used to describe diversification of income sources (see Ellis 1998: 30, n.4).
13. Barrett et al. (2001: 318) make a distinction between 'non-productive assets'

(e.g., household valuables), which directly generate 'unearned' income, and 'productive assets' (e.g., human or land capital), which produce income when they are allocated to activities such as farming or trade.

14. Barrett et al. suggest that the sectoral classifications of standard national accounting systems should be followed in order to maintain a logical correspondence between micro- and macro-level analyses (2001: 318).
15. Barrett et al. state that 'perhaps the most common error is classifying agricultural wage employment income as non-farm rather than as agricultural (sector) and off-farm (location)' (319).
16. See Gerand J. Gill (1991) for a review on seasonality and agriculture in the developing world.
17. Besides the economic literature, a wide social science literature has focused on understanding 'risk'. As Ryla Mehta et al. (1999: 12) mention, 'debates on risk have made a clear distinction between those understandings of risk that draw on a positivist understanding of science and those which focus more on the cultural and sociological aspects of risk'. According to Mary Douglas, 'every choice we make is beset with uncertainty. That is the basic condition of human knowledge. A great deal of risk analysis is about trying to turn uncertainties into probabilities' (Douglas 1985, in Mehta et al. 1999: 41, 1).
18. Covariate yield refers to yields that are similar across neighbouring households and/or regions (e.g., since local climatic conditions are similar).
19. For transaction costs, see section 2.4.3.
20. Economies of scope exist when the same inputs generate greater per-unit profits when spread across multiple outputs than when dedicated to only one single output. Economies of scope differ from economies of scale, in which per-unit profits increase as the amount of all inputs used for production goes up. Economies of scale tend to favour specialization (Barrett et al. 2001).
21. See also Jeannine Brutschin (2002) for a brief overview of the conceptualizations and uses of vulnerability.
22. Adaptation may be positive or negative: positive if it is by choice, reversible and increases security; negative if it is of necessity, irreversible and fails to reduce vulnerability (Davies and Hossain 1997, in Ellis 2000b: 14). Negative adaptation occurs when the poor can no longer cope with adverse shocks.
23. The World Bank (2002b: 37) also emphasizes the stability of institutions. However, institutions 'must be capable of changing and adapting, and new institutions must emerge'.
24. Scott, an organizational sociologist, identifies three elements which theorists can look at when dealing with institutions: a *regulative* system, a *normative* system and a *cultural-cognitive* system (Scott 1995: 51). These three elements

form a continuum moving 'from the conscious to the unconscious, from the legally enforced to the taken for granted' (Hoffman 1997, in Scott 1995: 51).

25. Markets means 'the way markets work in practice ('the markets' as an institution)' (North 1990, in Ellis 2000b).
26. The ten levels are the international level, the national level, the regional level, the district level, the subdistrict level, the locality level, the community level, the group level, the household level and the individual level (Uphoff 1993: 608).
27. Other studies on institutions have defined the levels that are influenced by institutions in a different way: de Haan and Zoomers (2005: 35) write, for example, of the micro, meso and macro levels.
28. On the critical role of informal institutions in developing countries, see the work of Hernando De Soto (1992), and for the specific Indian case Barbara Harriss-White (2003).
29. The notion of 'pull' (or positive) and 'push' (or negative) diversification has also been used by Ashley et al. (2003: no pagination).
30. As Ellis (1998: 23-5) highlights, the neglect of the determinants and effects of diversification differentiated between women and men is one of the weaknesses of using the household concept.
31. This definition is based on the work of Scoones and Wolmer (2002) in African crop-livestock systems.

CHAPTER 3

# Empirical Research: Framework and Methodology

This chapter details the empirical research implemented for this study. Section 3.1 highlights the overall research framework of the study, which formed part of a broader research programme. Section 3.2 specifies the research strategy underlying the empirical research, the methods used for data collection and analysis, as well as the selection of the case study area. The chapter ends with a discussion of some of the positive, as well as constraining, experiences during fieldwork.

## 3.1 RESEARCH FRAMEWORK

### 3.1.1 Overall Framework

The present study is part of the research programme 'Institutional change and livelihood strategies in South Asia' started in 2001 by the Development Study Group at the Department of Geography, University of Zurich. This research programme is embedded in the National Centre of Competence in Research North-South (NCCR North-South) and is entitled 'Research Partnerships for Mitigating Syndromes of Global Change'.[1] The NCCR North-South is a development research initiative which builds upon strong research partnerships between Swiss research institutions and institutions in developing and transition countries, so as to build up 'competence and capacity in order to develop socially robust knowledge for mitigating action' (SARPI 2000: 16). Between 2001 and 2005, the NCCR North-South consisted of eight individual projects (IPs), each with a specific regional and thematic focus. The Development Study Group at the University of Zurich is part of IP 6, entitled 'Institutional change and livelihood strategies', and has a research focus on Nepal, Pakistan and Kerala (Development Study Group 2001).

In Kerala, the research partner is the Centre for Development Studies (CDS) in Thiruvananthapuram. The CDS was previously involved in a collaborative research project on land-use dynamics in Kerala (Geiser et al. 1996).

An innovative programme component of the NCCR North-South is its Partnership Action for Mitigating Syndromes of Global Change (PAMS). PAMS are small-scale and time-limited development projects. They are demand-driven and implemented by partners (normally local NGOs) in developing or transition countries, and they have to show a strong linkage to research projects; a research partner also normally backstops them. Their main aim is to test strategies for mitigating the effects of global change on natural resources and human livelihoods (NCCR North-South 2002: 3). Two years after their initiation, PAMS have shown that they have a direct mitigation effect, and that they generate important lessons for research (Haupt and Müller-Böker 2005: 101). As such, they constitute an effective vehicle for knowledge and technology transfer.

### 3.1.2 Kerala Research Group

The aim of the specific research project in Kerala is to understand the impact of globalization on livelihoods in the marginal regions of the Western Ghats and to analyse the threats and opportunities of economic globalization and state decentralization (Müller-Böker 2004: 255). The main hypothesis is that processes of accelerated global interdependence impinge in an uneven manner on nation-states and on various sectors and regions within states. A further hypothesis is that, at the local level, institutional arrangements play a key role regarding people's socio-economic mobility, both upwards and downwards, and that the process of globalization renders these local institutions more dynamic and sensitive.

The Kerala Research Group consisted of scholars working on several sub-projects. These research groups embarked upon fieldwork either by selecting a specific village or ward as a case study or by studying a specific sector (e.g., the coir or rubber sector). The local research partners were based at the Centre for Development Studies (CDS), under the guidance of Prof. K.N. Nair. Between 2002 and 2005, several workshops took place at the CDS to facilitate exchanges about conceptual and methodological issues and to present preliminary findings. Two overview papers were written by members of the research group; one on the impact of the WTO on Kerala (Nair and Menon 2004) and one on 'Understanding the globalization discourse' (Backhaus 2003).

### 3.1.3 Natural Rubber Sub-Project

The present sub-project was accompanied by two Masters theses and a *partnership action* (PAMS) focusing on different aspects of the natural rubber sector in Kerala. A first thesis by Zollinger (2005) took a gender-specific perspective and analysed the impact of fluctuating natural rubber prices on the income-generating and expenditure strategies of women-headed rubber holdings in Kerala. The second Masters thesis (Moktar 2006) studied the potential and the limitations of introducing a fair trade scheme for natural rubber produced by marginal rubber holdings in Kerala. This work analysed the natural rubber value chain and the views of the different stakeholders concerning fair trade.

The PAMS project that accompanied the sub-project had the aim of testing and evaluating an IT-led marketing system for vanilla.[2] In fact, vanilla has shown to be a crop which can occasionally make a major contribution to the income of certain rubber holdings in Kerala. But farmers growing vanilla had difficulties in getting a fair price for their product and wished to become organized and build up a common marketing platform, sustained by a professional IT database. The project took place between March 2004 and April 2005 and was implemented by INFACT (Information for Action), a local NGO which has worked for sometime with rubber planters.

## 3.2 METHODOLOGY[3]

### 3.2.1 Research Strategy

Case study research was adopted as the core strategy to investigate income generation and the local institutions. The choice of case study research has proven useful for the several reasons. Firstly, this research strategy allows investigation of a contemporary phenomenon, such as income generation, in its real-life context, and therefore does not seek to separate the phenomenon from its context. The case study is thus holistic, consisting of facts gathered from various sources and conclusions drawn from these facts (Tellis 1997: no pagination). When he stresses the importance of the (local) context, Véron is right in stating that 'it would be more useful to expand theoretically informed empirical research of actual problems . . . in specific contexts. Such a research programme would involve due consideration of local diversity and of locality-specific solutions' (Véron 1999: 182). Secondly, the advantage of case studies is that they can be generalized in the sense that they aim to expand theoretical propositions instead of aiming at representing 'samples' which can then be generalized to larger populations or universes. They are thus analytically rather than statistically generalizable (Yin 2003: 10). Thirdly, case studies are a suitable strategy when asking questions of 'how' and 'why'. Furthermore, employing the case study approach is appropriate in situations where the investigator has little control over events (e.g., the behaviour of people cannot be manipulated) and when a current event can be studied by interviewing the people involved (in contrast to historical studies, where this is not often the case). For all these reasons, the case study approach has proved useful in understanding the complex social phenomena under study in this research project.[4]

The case study design chosen was the *embedded case study* (Yin 2003: 42), which occurs when, within a single case, attention is given to sub-units. This allows the case study to involve more than one unit of analysis within the same context, such as, for example, the different types of rubber holdings. The case study research also furnished a methodologically sound conceptualization of the comparative framework.

### 3.2.2 Methods and Data Collection

Field research was done using a combination of qualitative and quantitative methods which had already shown itself to be useful during a previous research project (Strasser 2000).[5] In fact, there are many advantages to using both qualitative and quantitative methods. Firstly, a triangulation of methods permits the examination of many different aspects of the study subject over a relatively short period of time. Contradictions can thus be revealed in a way which is not possible using a single method. Backhaus correctly states that using many different methods can make sense within the wider context of problems human geography is confronted with, because broadening the study is often more interesting than going into greater detail (Backhaus 2001: 951). White, in his paper about the combination of quantitative and qualitative approaches, puts it as follows:

> Both quantitative and qualitative techniques have their place in social analysis. There is no reason to give primacy to one over the other. Different methods are required to tackle different problems, and a combination of techniques will frequently yield greater insight than other one used in isolation. (White 2002: 519)

On the subject of combining methods within case study research, Yin notes that it should not be confused with qualitative research and can well include quantitative evidence (Yin 2003: 14).

The present study is based on primary data collection over a period of twelve months in Kerala during three field visits between March 2003 and March 2005. The study also draws on secondary data from official statistics, government reports, newspapers, journals, published books and—with rapidly increasing importance—the Internet. The whole research process was carried out in four phases as is sketched out in Fig. 3.1. The research process has, however, not been quite as linear as it appears in the figure. In fact, it has been more of an iterative process,

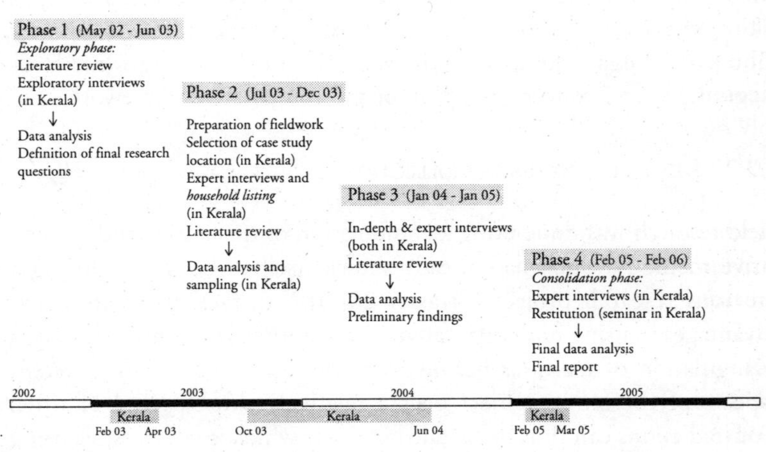


*Source:* Author's figure.

Fig. 3.1: Main data collection phases.

where the predefined workplan often had to be revised in order to readapt research questions and methodology to the reality of the field. My experiences support those of Backhaus and Steinemann (2001: 15), who write that working out a good research question is one of the most demanding tasks in a research project.

The first data collection phase was exploratory. It started with a literature review in Switzerland and was followed by a two and a half month-long field visit to Kerala from February to April 2003. During the field visit, exploratory and unstructured interviews (based on a snowball sampling) were carried out with eighteen stakeholders involved in production, processing and marketing, as well as research and administration, in the natural rubber sector. The main aim was to get a broad overview and understanding of the current challenges facing the rubber sector and to find a suitable location for the case study. In addition, local literature was collected and reviewed. The key research questions were refined and the field methods adapted following analysis of this material.

The longest field visit was in the period from October 2003 to June 2004, during which the core data was collected. The second data collection phase was after the selection of the final study location. In December 2003, three college students conducted a *household listing*[6] amongst the 299 households in the study location, based on the electoral roll of the Panchayat elections of 2000.[7] The aim of the household listing was to obtain a list of all households in the location, including some of their characteristics (e.g., household size, land ownership, main income pattern). The household listing proved to be of great help in solving what Gerritsen has called the 'fuzzy empirical situation', i.e. obtaining samples is problematic in those areas where data on the population to be studied is unreliable or even absent, as is often the case in the developing world (Gerritsen 2003: 3). This quantitative data provided the framework for raising the questions to be addressed in the subsequent qualitative approach. Based on this list of households, a stratified random sampling was done of the rubber holdings (i.e. those holdings who had not planted rubber after 1993 were excluded from the study).

The third data collection phase consisted of *semi-structured, in-depth interviews* conducted with forty rubber holdings and around fifteen local actors such as rubber dealers, agricultural commodity traders, agricultural officers, Panchayat members, bank managers, and many others. These interviews—which generally lasted between 1½ and 3 hours—were conducted during the period from January to May 2004. A semi-structured questionnaire ensured that all the core issues were addressed, so as to enable comparison. During interviews with rubber holders, questions were asked about their income and coping strategies, as well as their linkages to local institutions and organizations. Furthermore, quantitative data was collected to assess the gross annual income of the farm, and a short historical farm profile was done. After each farm interview, a visit to the informant's plot (*observation*) served to verify and clarify oral information (*triangulation*). This type of interview (*in-depth, semi-structured*) allowed for flexibility and openness, since the interviewee was able (and welcome) to speak about what he/she felt was important. Consequently, many interviews threw up new and important topics.

The fourth and last data collection phase took place during a final trip to Kerala in March 2005. It was mainly a consolidation phase. Some final interviews were conducted with expert farmers to cross-check and triangulate some of the data. More importantly, this last journey also provided the possibility for the restitution of first results. Indeed, I was invited to present some preliminary results at a seminar organized by the Rubber Research Institute of India, and this proved an important step, both for sharing my knowledge and views and for getting some feedback from the participants. Results were also presented to a small group of rubber planters from the region.

Data analysis took more time than initially anticipated. Firstly, one is often surprised once back in office at the sheer amount of interesting field data collected, and secondly, data analysis is an important iterative 'trial-and-error' process which needs—and deserves time. As Rubin and Rubin usefully state, 'Data analysis is the final stage of listening to hear the meaning of what is said' (Rubin and Rubin 1995). Also, data analysis forces one to forget one's preconceived views about how local reality functions. This can only be done with a strict use of data analysis techniques; this is essential to ensure the full integrity of one's results. The analysis of the qualitative data was conducted following Silvermann (2000) and using the ATLAS.ti software package. The surveyed data from the household listing was analysed using the CSPro software package.[8] Income data gathered during interviews was analysed using a standard spreadsheet application.

### 3.2.3 Selection of Study Location

It took several attempts to select a suitable study location. Initially, the locality of Meenachil Panchayat (Panchayat refers to the local government) near the city of Palai in the Kottayam district was selected. However, the staff of the Rubber Research Institute suggested several locations which were more suitable for this research due to their agro-climatic conditions less favourable for rubber cultivation. Exploratory visits, as well as transect walks and drives[9] through these locations led to the decision to work in Thalanadu Panchayat of Kottayam district in the higher midlands of central Kerala (Fig. 3.2).

Besides the marginality of the location, there were other reasons as well that determined in the choice of this Panchayat. Despite constituting one administrative unit, Thalanadu is made up of areas which demonstrate very different characteristics. The choice of two wards, Choovoor and Maravikkallu, with very different social and geographical patterns provided an opportunity to set-up a comparative framework. Last but not the least, their (relative) proximity to the field office, located in the town of Palai, allowed for the interviews to be conducted during the day, typed up in the late afternoon and evening whilst still fresh in the memory.

### 3.2.4 Experiences with Methodology and Methods

The field study was a rich experience. The almost daily encounters with unexpected situations or events and interactions with people from a different culture and background make field research an exciting, though demanding endeavour. Some of the positive experiences, as well as the hurdles, are worth mentioning specifically.

*Doing in-depth interviews*

Qualitative interviewing had already proved a fascinating adventure during previous research projects, every step of an interview bringing up new information and opening windows into people's experience. The experience in Kerala was similar. The research team was almost always welcomed at farm houses and people were curious about the purpose of the visit. In most cases, interviews could be done immediately; in some

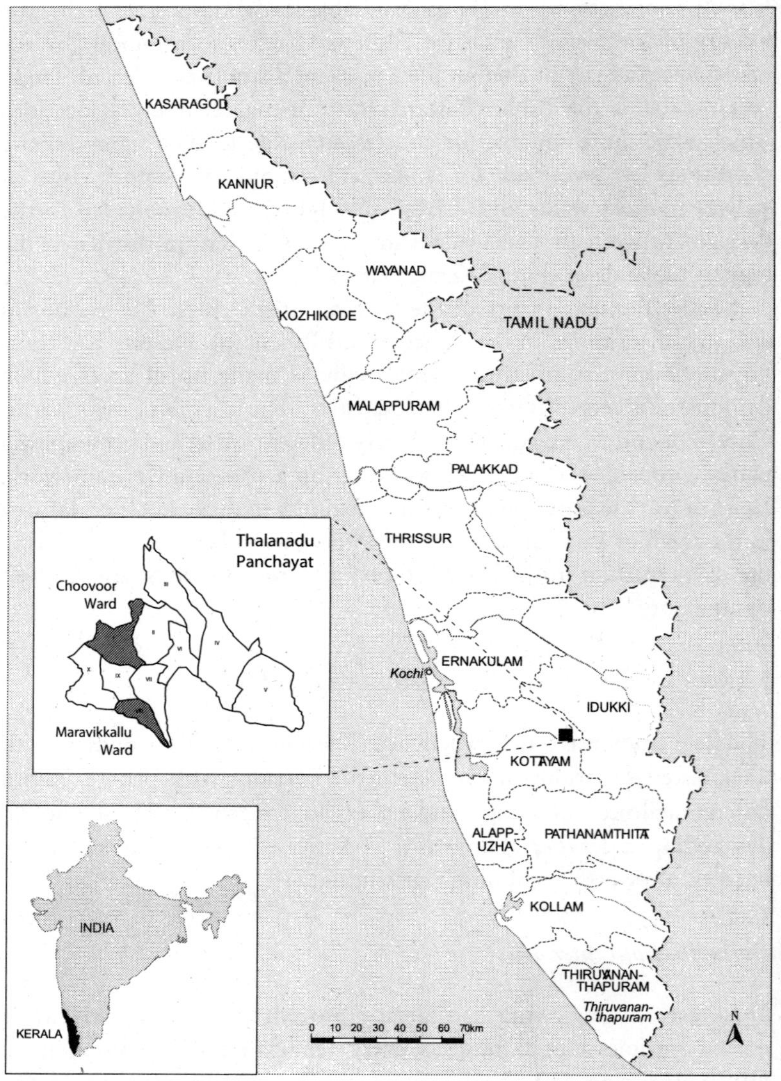


*Source:* Author's figure (graphic design by Martin Steinmann).

Fig. 3.2: Kerala and the study location.

other cases, a suitable time during the following days was agreed upon. In contrast, it was more difficult to access and interview staff from private companies or government administrations, though the latter could finally be interviewed once the necessary permits had been obtained.

One difficulty was to get female interviewees to answer questions. In some cases, interviews might have been started with a woman on her own, but during the interviews, male family members or neighbours (sometimes members from the extended family, sometimes not) showed up and took over answering the questions. It was therefore difficult to adequately represent women's viewpoints, though they did answer some questions which specifically concerned their income-generating activities. However, on gender aspects, the present research builds on results presented in the Masters thesis of Zollinger (2005) (see section 3.1.3).

*Working with an interpreter*

Malayalam is the predominant language amongst farmers in the study location, and very few interviewees had any knowledge of the English language. It was therefore essential to work with a local interpreter who translated conversations directly into English. This was one of the major constraints during the interviews since it made it difficult to react promptly to newly emerging issues or to directly interpret what was 'between the lines', normally not translated. However, the main difficulty in working with an interpreter is guaranteeing good quality translation. Imprecision and distortions were inevitable, but this was compensated for by two interpreters with a great ability to motivate and entertain the interviewees. They were adept at establishing a nice, sometimes jokey atmosphere right from the beginning of an interview. Because this built trust, it often facilitated getting very interesting in-depth information. Besides this, the interpreters' knowledge of the local area and customs also helped in organizing the interviews.

*Research for development?*

Whenever possible, research should contribute to sustainable development (KFPE 2003: 26). It is especially favoured, in the context of the NCCR North-South where research is specifically aimed at the 'mitigation of syndromes' (section 3.1.1), to overestimate the potential contribution of one's own research to development. My experience has shown me that one has to be realistic in assessing where and how results can be applied, and that a mixture of development research and development practice can, unexpectedly, become problematic. I'll use an example from my own research to illustrate this.

As I mentioned at the beginning of this chapter, I worked together with a local NGO on the planning and implementation of a partnership action (PAMS) which was linked, to a certain extent, to my research questions. Though it had not been planned that way, I soon realized that it was necessary to make a clear separation between my case study location and the development project area in order to be able to implement my research activities properly. Not to have done this would undoubtedly have affected my reputation as an independent researcher and possibly led to biased answers. From that moment on, I took great care not to mix research matters with project matters. My apprehensions were confirmed by the experience of a colleague doing research in Delhi who implemented a PAMS towards the end of her studies. After the start of the PAMS, she was suddenly looked upon as a donor and no longer as a researcher. One can imagine the consequences this can have for sound research.[10]

In conclusion, I argue that a well-planned restitution and dissemination process is more important than practical activities in the field. Proper restitution and dissemination of results should not be limited only to the scientific community and the research partners; it is of the utmost importance that the results be presented to the people and institutions that can finally use them. All too often, this is planned at the beginning but not actually done at the end of the research, when time and financial resources have already been almost exhausted.[11]

### *Local collaboration*

One positive aspect linked to the implementation of the PAMS must be mentioned here. Through this partnership action, I entered into a close and very regular collaboration with INFACT, the implementing NGO. This collaboration was instrumental in helping me to set-up a good fieldwork situation, as well as to locate and gain access to the study location. In addition, staff members became my occasional field assistants and helped with the organization and translation of interviews and meetings. Ultimately, this grew into the type of partnership that is so wonderful to experience as a researcher in an unknown context.

## NOTES

1. The main stated research objective of the NCCR North-South is to contribute to an improved understanding of the status of different syndromes of global change, the pressures these syndromes and their causes exert on different resources (human, natural, economic), and the responses of different social groups and of society as a whole (SARPI 2000: 16). The NCCR North-South is financed by the Swiss National Science Foundation (SNF) and the Swiss Agency for Development and Cooperation (SDC).
2. Further details on the project can be downloaded from the NCCR North-South website: http://www.nccr-north-south.unibe.ch (accessed August 2005).
3. In the literature, terms such as 'methodology' and 'methods' are often used synonymously. Here, a distinction is made: the term methodology refers to the 'rules' that researchers have to follow when designing and conducting scientific research. These rules indicate the way in which a study is to be performed in order to obtain valid answers to the research objectives and questions (Gerritsen 2003: 1). As Groot states: 'a methodology shapes and structures the learning process' (1998: 2). A methodology contains a number of methods (e.g., transect walk) in a particular sequence.
4. On the need for case study research, refer to Flyvberg (2004).
5. For a full justification of the applied methods, refer to Flick (1998), Marshall (1994), Mason (2003), Silvermann (2000) and Mikkelsen (1995) for the qualitative research, and to Rubin and Rubin (1995) for 'the art of interviewing'.
6. A ***household listing*** is a short, focused, quantitative survey. Its main aim is to get an overview of all the households present in the study location. A further sampling can then be based on this household list and on some specific characteristics of the households (Strasser 2004).
7. The availability of and access to the voters' list was a great help in organizing the ***household listing*** though some adjustments has to be made to the list due to outdated data.
8. CSPro (Census and Survey Processing System) is a public-domain software package for entering, editing, tabulating and mapping census and survey data. It can be downloaded under http://www.census.gov/ipc/www/cspro/. For a tutorial, see U.S. Census Bureau (2003).
9. Transect drives are nothing more than drives through the main roads of a location including the neighbouring areas. Having a rough overview over a

location and its surroundings is in fact very useful before getting more into the field, for example, with a transect walk.

10. Personal communication from Susan Thieme, July 2004.
11. On the importance of restitution/dissemination, see the excellent booklet titled *Guidelines for Research Partnership with Developing Countries* (KFPE 2003). Dissemination of results is one of the eleven principles for a sound research partnership (KFPE 2003: 24).

CHAPTER 4

# Rubber in Kerala

## 4.1 KERALA: ITS DEVELOPMENT AND THE ROLE OF AGRICULTURE

### 4.1.1 A Profile of Kerala[1]

*Location and biophysical features*

The state of Kerala[2] is situated in the south-west of the Indian subcontinent. It came into existence in 1956 through the merger of the three former administrative units of Travancore (in the south), Cochin (in the centre) and Malabar (in the north). Malayalam is the state language, spoken only in Kerala.

Kerala has an area of 38,864 sq km (roughly the size of Switzerland) and it lies between 8° and 13° north latitude and 74° and 77° east longitude. It is confined to the east by the Western Ghats, to the west by the Arabian Sea, to the south by the State of Tamil Nadu, and to the north by the state of Karnataka.[3] Kerala is characterized by an irregular topography and its altitude ranges from below mean sea level to 2,694 metres above mean sea level. It is usually divided into three distinct elevation zones, known as the lowland, midland, and highland regions.[4] The lowland region is a strip of land running along the 580 km-long coast of the Arabian Sea. It varies from nearly level to gently-sloping. The lowland region occupies approximately 10 per cent of the total area and is well known for its backwaters, which are fringed with extensive rice fields and coconut plantations. The midland region accounts for 60 per cent of Kerala's total area. It is mainly made up of flat-bottomed valleys and gently to moderately sloping areas with small hills, and numerous rivers. A variety of seasonal, annual and perennial crops are grown in its mainly lateritic, moderately fertile soil. The highland region

—which accounts for 30 per cent of Kerala's area—lies in the east of the state and is part of the Western Ghats. It includes steep, mountainous terrain but also fairly even, high plateaux. The region is covered with forests and drained by small streams. Plantation crops, mainly tea, coffee and cardamom, are grown in the highlands. Kerala's location and altitudinal variations have endowed the state the rare position of having a wide range of agro-ecological conditions.[5]

Kerala's climate can be characterized as tropical, i.e. hot and humid. The mean rainfall, at approximately 2,900 mm, is very high. About 90 per cent of the rain falls during the two monsoon seasons. The south-west monsoons bring rainfall from June to September and the north-west monsoons in October and November. Generally speaking, the northern districts have the longest and most pronounced dry season, but they also experience massive rainfall during June and July. In central and southern Kerala, the rainfall is more equally distributed.[6] During the period 1956-93, the mean annual temperature varied from 25.4 °C to 31 °C in the midland region. The temperature was lower in the upper highlands, sometimes falling below 15 °C in the north-eastern district of Idukki.

### *Demography and administration*

At the time of the 2001 Census of India, Kerala had 31.84 million inhabitants. This represents 3.1 per cent of the total Indian population. The population density is 819 inhabitants per sq km, one of the highest in India (mean density: 324), as well as the world (Switzerland: 184, China: 136). Urban areas account for 26 per cent of the population, while 74 per cent of the population lives in rural areas. However, in Kerala, the distinction between urban and rural is problematic due to the unique settlement pattern. In the rural areas the population does not live in clustered villages but on scattered homesteads. These areas are very densely populated and their boundaries with the few interspersed small towns are unclear (Véron 1998: 71-2). This specific settlement pattern is often called *rurban* and its developmental process *rurbanization* (Pronk 1997: 21-2).

For administrative purposes, the state is divided into 14 districts,[7] 152 community development blocks, 991 village Panchayats (local political-administrative authorities in rural areas), 53 municipalities (town areas) and 5 corporations[8] (cities). Panchayats, municipalities and corporations are further subdivided into wards. The Panchayats are governed by elected presidents and executive members, and administered by government employees. Alongside this political-administrative territorial division, there is a separate system for revenue collection. The 14 districts are divided into 63 *taluks* and 1,452 revenue villages. The borders of revenue village and of Panchayats are not identical, but they often more or less coincide.

### *Religion and caste*

The religious composition of Kerala is more balanced than that of India as a whole. According to the 2001 Census, 56.3 per cent of the population is Hindu, 24.7 per cent is Muslim and 19.0 per cent is Christian. Compared to the 1981 Census of India, the proportions of Hindus and Christians in the total population declined slightly, while that of Muslims increased in the interim period. Despite still having a very rigid caste system in the early twentieth century, Kerala today is one of the least caste-ridden societies in India. The following list[9] gives a rough impression of the most important groups in contemporary Kerala and the main income activity with which they are traditionally linked:[10] the *Namboodiri* (Brahmin elite, the top of the social scale), the *Nair* (heterogeneous activities: farming and rural industries, but mainly in state administration), the *Ezhava* (cultivation and exploitation of the coconut palm), the *Pulayas* (mainly agricultural labourers), and the *Adivasi* (tribal populations, not a caste *stricto sensu*). Nowadays, the Constitution of India gives special status to 'outcaste' communities, which are divided into *Scheduled Castes* (mainly Dalits) and *Scheduled Tribes* (mainly Adivasis), commonly abbreviated as SC and ST respectively. These communities are assisted in various ways, including having seats and jobs in government reserved for them. In Kerala, SCs and STs respectively, account for 9.8 per cent and 1.1 per cent of the total population.[11]

The proportion of Christians in Kerala is high compared to the national average.[12] This is due, amongst other things, to the numbers of Christians in the central districts of Kerala, which is also the main area of rubber cultivation. The most important group are the Syrian Christians, who have been instrumental in developing the rubber plantations. Having first joined British companies as workers, supervisors and agents, they diversified into moneylending, industry and trade. Subsequently, they established credit institutions and modern banking on the one hand, and vertically integrated agribusinesses, such as rubber and tea plantations, on the other. Syrian Christians today are well-represented in management, both in the state and in the corporate sectors (Harriss-White 2003: 152-3).

### 4.1.2 Development in Kerala

Kerala has become well known for its unique development pattern. Despite showing poor economic indicators over long periods, the state displays a set of very high social (and human) indicators of development which stand out in comparison to the rest of India. Briefly, some of Kerala's most remarkable social development characteristics are a life expectancy of 70 years (59 for the whole of India), an infant mortality rate of 11 per 1,000 live births (72 for India), and a literacy rate of 90.9 per cent (65.4 per cent for the rest of India).[13] These indicators are only slightly lower than those of industrialized countries. Until the 1990s, however, the advances in the field of social development failed to induce economic development. Kerala's per capita Net State Domestic Product (NSDP) of about Rs. 1,508 (in 1980) to Rs. 3,718 (in 1989) and an average yearly growth rate of 1.06 per cent during the 1980s were very low and below the Indian average (all in current prices).[14] This unique development experience of Kerala, displaying the paradox of high social development and economic backwardness, is often referred to as the 'Kerala model of development'. This 'model' has led to a series of studies and debates (Prakash 1994; Franke and Chasin 1998; Véron 2001). However, recent studies have shown that a turnaround in economic growth has taken place during the 1990s (Chakraborty 2005; Kannan 2005). Some argue that this has been spurred by the economic reforms implemented by the central government in the early 1990s, which had a substantial impact on Kerala's economy (Prakash 1999: 27). Although Kerala's economy could not sustain the tempo of such development after the mid-1990s (ibid.), it is a fact that the state's economy has shown an impressive growth performance in the period from 1987-8 to 2002-3, with a growth rate of the NSDP of 5.99 per cent during the 1990s.[15] This is the highest growth rate since the creation of the state in 1956. Thus, 'Kerala appears to have gone past the syndrome of high social development and low economic growth' (Kannan 2005: 549). Nonetheless, key issues raised in many debates—such as the fiscal crisis of the state in Kerala; the continuing high unemployment, particularly amongst the educated; the slow but steady grip of a culture of rent-seeking; the contradictory position of the organized working class in opposing technological change while successfully bargaining for higher wages and other benefits; and the inability of the state to attract adequate investment due to its negative image as an 'investor-unfriendly State'—remain important and unsolved (548).

### 4.1.3 Agriculture in Kerala: Importance, Significant Changes and Trends

Due to its favourable agro-climatic conditions and physical location, agriculture has long been an important pillar of Kerala's economy, which today depends to a great extent on agriculture, both for income and employment, though during the last years, there has been a relatively slow growth of the sector—a situation similar to the rest of India (Kannan 2005: 549). In 2000-1, the primary sector[16] accounted for 27.4 per cent of the Net State Domestic Product compared to 24.1 per cent for the secondary and 48.5 per cent for the tertiary sectors (Table 4.1). Employment-wise, the primary sector accounted for 32 per cent of total employment, compared to 28 per cent for the secondary sector and 40 per cent for the tertiary sector (various sources in Kannan 2005: 552).

Kerala's agriculture has, however, undergone important changes since the 1950s, especially concerning shifts in cropping patterns and the consequent change in land use. In the first half of the twentieth century, the cropping pattern was predominantly based on traditional food crops,

TABLE 4.1: DISTRIBUTION OF NET STATE DOMESTIC PRODUCT (1980-1 TO 2000-1)

| *Sector* | *1980-1* | *1990-1* | *2000-1* |
|---|---|---|---|
| Primary | 39.23 | 35.94 | 27.4 |
| *from which agriculture* | *33.85* | *33.43* | *24.9* |
| Secondary | 24.37 | 24.02 | 24.1 |
| Tertiary | 36.40 | 40.04 | 48.5 |

*Sources*: For 1980-1 and 1990-1: Kerala State Planning Board (1994, 1997), cited in Prakash (1999: 33); for 2000-1: Kerala State Planning Board (2002).

mainly rice and tapioca (the main cereal substitute). During the 1950s and 1960s, the newly formed state government continued with its policy of promoting rice cultivation, but since the mid-1950s, a gradual shift in the cropping pattern took place towards tree crops, especially coconut and rubber (Nair and Menon 2004). The year 1975 is considered a turning point in cropping patterns, as it was at this moment that the area under rice cultivation reached its peak. After 1975, there was a clear shift away from food crops (including tapioca) towards tree crops, such as rubber and coconut, and export-oriented crops such as pepper, ginger and coffee (George and Chattopadhyay 2001: 88). Table 4.2 illustrates this dramatic change in cropping patterns between 1957 and 2001-02, the area of rice cultivation declined by 58 per cent, while the areas planted with coconut and rubber increased by 95.7 per cent and 475 per cent respectively (aggregated data from Table 4.2), both surpassing rice in terms of hectarage.

These shifts have significant implications for the food security of the state, which already depends on outside supplies to meet more than half of its food requirements (George and Chattopadhyay 2001: 88). In fact, rice production declined by almost 50 per cent between 1975 and

TABLE 4.2: CHANGES OF CROPPING PATTERN IN KERALA 1957-2001/2

| | *Area (in 1,000 ha)* | | | *Per cent change* | |
|---|---|---|---|---|---|
| | *1957* | *1975* | *2001/2* | *1957-75* | *1975-2001/2* |
| Total cropped area | 2,211 | 2,933 | 2,992 | 32.6 | 2.0 |
| Rice | 767 | 876 | 322 | 14.2 | -63.2 |
| Other cereals/pulses | 28 | 45 | 29* | 60.7 | -35.6* |
| Pepper | 91 | 108 | 204 | 18.7 | 88.9 |
| Cardamom | 28 | 54 | 41 | 92.9 | -24.1 |
| Banana and plantain | 42 | 52 | 107 | 23.8 | 105.8 |
| Tapioca | 214 | 327 | 112 | 52.8 | -65.7 |
| Coconut | 463 | 693 | 906 | 49.7 | 30.7 |
| Coffee | 17 | 42 | 85 | 147.0 | 102.4 |
| Rubber | 100 | 207 | 475 | 107.0 | 129.5 |

*Source*: *Economic Review*, Kerala State Planning Board (various issues). Adapted from George and Chattopadhyay (2001: 88).

*Note:* *The figures for 'other' cereals/pulses in '2001/2' are from 1996.

2000-1, from approximately 1,400 thousand tonnes to 703 thousand tonnes (Kerala State Planning Board 2002: 41). Apart from the food security issue, the irreversible conversion of rice lands into agricultural land used for other crops (as well as for non-agricultural purposes)[17] has serious ecological as well as socio-cultural implications (George and Chattopadhyay 2001: 90-1; Portmann and Strasser 2005). There are multiple reasons for this switch in cropping patterns. The increased cost of cultivation and reduced profitability are seen as the major factors influencing the shift from rice to other crops; increased costs are mainly due to the increase in wage rates of agricultural labourers[18] (Santhakumar and Nair 1999: 316). Moreover, state policies on land, labour and taxes[19] created an environment which promoted the plantation of commercial crops (George and Chattopadhyay 2001). In 2001, 29 per cent of the cropped area of the state was planted with the four important commercial plantation crops, namely rubber, tea, coffee and cardamom. Moreover, 43 per cent of the total area under these crops in India is located in Kerala, showing the importance of plantation crops in the state (Kerala State Planning Board 2003: 45). The consequences of this cropping pattern shift are that perennial and tree crops have brought greater rigidity into the patterns of agricultural growth since they limit the options of farmers to change their crops in response to short-term fluctuations in prices (Nair and Menon 2004).

The limited possibility for changing crops in the short term became clear during the plantation sector crisis, which started at the end of the 1990s and, for certain crops, continues even today. In fact, with the withdrawal of *quantitative restrictions* (QRs) on imports of agricultural commodities as of 31 March 2001, the crops in the state which has benefit from protection by the Indian government became vulnerable to competition from imports. Although the quantity of imports as a percentage of domestic production is relatively low, they do affect domestic prices due to different mechanisms (Nair and Menon 2004: 25-6).[20] Recent local studies have shown the negative influence of low commodity prices on smallholders in rural Kerala (Mahesh and Vinod 2005; Ramakumar et al. 2005), resulting sometimes in farmers committing suicide (NZZ 2002).[21] However, the price factor is only one component of the agrarian crisis, another important component being the structural constraints affecting the sector, some of which are the proliferation of small and marginal holdings due to subdivision and fragmentation of land; the tremendous increase in land value with the effect that people have started viewing land as a speculative asset and not as a factor of production; the fact that the share of people with non-agricultural interests owning agricultural land (so-called gentlemen farmers) is constantly increasing; the lack of effective water management and ecological consequences of poor land management; the lack of appropriate research systems for the different crops; the lack of adoption of agricultural practices which could increase productivity; and a generally weak agricultural extension system (apart from rubber, see section 4.3) (Nair and Menon 2004: 28-32).

## 4.2 THE CONTEXT OF RUBBER CULTIVATION

### 4.2.1 Botany, Ecology and Management[22]

*Hevea brasiliensis*, the rubber tree, is the most important commercial source of natural rubber, a product recovered from its latex.[23] *Hevea brasiliensis* belongs to the family of the *Euphorbiaceae*. It is essentially a tropical plant native to the tropical rainforests of the Amazon Basin in South America. It grows mainly within a well-defined rubber belt 1,126 km north and south of the Equator, and the main growing regions are the tropical regions of Asia, Africa and America (Fig. 4.1). However, the development of new varieties has made the plantation of rubber possible outside the rubber belt. *Hevea* is a tall tree attaining a height of about

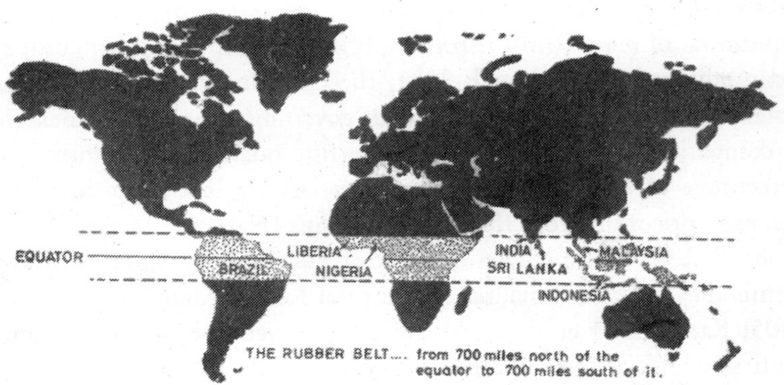


*Source:* Kochhar (1998: 407).

Fig. 4.1: World map showing the rubber belt.

20 m; the trunk is 0.5 to 1.5 m in girth (older varieties having a larger girth) and bears a spreading or conical leaf canopy at its top. Rubber thrives in the lowlands (up to 300 m above msl.) but it can grow up to an elevation of 1,000 m above msl. The optimum agro-climatic requirements include a fairly evenly distributed rainfall of 175-250 cm per year,[24] a temperature range of 24 to 35 °C, a high atmospheric humidity of the order of 80 per cent, and a well-drained, fairly deep, loamy soil with a pH value of 4.5-6.0. Rubber can, however, also be grown successfully under moderately deviating conditions.

*Hevea* seeds are viable for only a short time and must therefore be planted immediately after harvesting. All earlier plantations in South and South-East Asia were raised from unselected seeds, the results being a low yield potential and consequently a low production. Selection work on *Hevea brasiliensis* to improve the planting materials and the introduction of vegetative propagation by budding led, over time, to the establishment of numerous valuable high-yielding clones. The new trees have a vigorous rootstock from one parent, a highly productive trunk from another, and a disease-resistant as well as photosynthetically efficient top or crown from a third parent. Varieties are selected for their disease resistance, their ability to retain foliage in the dry season, their resistance to wind damage and their yield potential. Systematic breeding work in rubber has resulted in the selection of varieties with potential yields of over 4,000 kg per ha.[25] in comparison with the 300-400 kg per hactare obtained from plantations raised from progenies of the Wickham trees.[26]

The trees are ready for tapping when they are about six to seven years old. At the first tapping, only a small amount of viscous latex exudes, but the yield increases progressively and the tree reaches full productivity at the age of twelve. The trees are cut down after twenty five to thirty years, when latex collection is no longer economically profitable. Tapping, the process leading to the collection of latex, has greatly evolved over time.[27] Nowadays, trees are normally tapped using the spiral panel system. In this system, a typical cut at an angle of 30-5° is made from the upper left to the lower right, halfway around the tree. A specially designed knife made of high quality steel is used. The knife has a V-shaped head which can be adjusted to cut the proper thickness of bark. The best yield is obtained by tapping to a depth of less than 1 mm close to the cambium, since this is where the greatest number of latex vessels are concentrated.

Fig. 4.2. Rubber plantation in Kottayam district, Kerala (24 April 2004).

Tapping begins in the early morning hours, when the flow of latex is copious owing to the high pressure in the tree's cells. The flow slows down as time passes and finally stops around noon. Latex runs down the channel of the cut to a spout and into a small receptacle. The yield of latex can be increased, sometimes by as much as 30 per cent, by applying chemical growth hormones just below the tapping cut. Before each tapping, coagulated latex on the cut is removed by hand and kept separate as scrap. Scrap rubber is sold and processed into rubber of inferior quality. Trees are normally tapped every second day (known as 'd/2'), but there are other tapping systems (called low frequency tapping) in which trees are tapped less often, every third (d/3), fourth (d/4), or even sixth day (d/6). Theoretically, rubber trees can be tapped the whole year round if rain guards are applied above the tapping cut during the rainy season to prevent water from flowing into the latex. However, in south India rubber trees shed their leaves from December to February; during this time the yield is comparatively poor and tapping can become uneconomical.

Rubber is collected from plantations in the form of latex[28] and scrap rubber (also called field coagulum). Latex and field coagula are highly susceptible to bacterial action and it is therefore essential to process them into forms which will allow for safe storage and marketing.[29] Processing is done on-farm or in processing centres. Latex is coagulated and then sheeted by passing it through a set of smooth rollers. The rubber sheets are then dried (Fig. 4.3) and smoked, either in a smokehouse or above

Fig. 4.3: Rubber sheets drying in the sun (12 January 2004).

an open fire, for example in a kitchen. The substances present in the smoke prevent the growth of mould on the smoked sheets. Smoked sheets are known as ribbed smoked sheets (RSS). Nowadays there is also a good market for preserved field latex and latex concentrates, as they are an important raw material with a wide range of applications. The choice of the processing method depends largely on economic considerations, the investment capacity, the availability of technical manpower, as well as on market demand (Rubber Board 2004: 33).

### 4.2.2 History, Production and International Trade

The Incas of Peru, the Aztecs of Mexico and the Maya of Central America used natural rubber from *Hevea brasiliensis* even before the 'discovery' of America by Columbus. Only in 1736 did the French physician M. de la Condamine despatch samples of the mysterious plant 'caoutchouc' (a French rendition of the Caribbean word *cahuchu*, meaning weeping tree) to France. He included a complete description of the tree, and the methods of collection and processing—and thereby stimulated the demand for rubber. The sample was obtained from a local tree called 'heve'; this name was later used to give the generic name *Hevea*. The year 1770 is memorable in the history of natural rubber because it was reported that hardened latex (*caoutchouc*) had the ability to erase pencil marks on paper. This property led to *caoutchouc* becoming known as 'rubber'.

Further important developments were made in the nineteenth century. MacIntosh discovered that rubber dissolved in naphtha offered a new way for producing waterproof articles such as raincoats. Another significant achievement was the development of the vulcanization[30] process by an American rubber manufacturer, Charles Goodyear, in 1839, which revolutionized the rubber industry overnight. Following his invention, factories for manufacturing rubber products began to spring up in the principal industrial countries. The rubber tyres used on roads were initially solid, but the new industry received a great boost when John Dunlop, an Englishman, developed the pneumatic bicycle tyre, from which the motor-car tyre was later developed (Kochhar 1998: 402).

As the nineteenth century drew to a close, the world's supply of natural rubber still came almost entirely from wild trees from the Amazonian rainforest. However, it soon became difficult to obtain enough raw rubber from Brazil to supply the rapidly growing industries of the industrial nations. As a consequence, many attempts were made to establish a more stable source of supply. In 1876, Henry Wickham managed to smuggle and ship 70,000 rubber seeds from Brazil to Kew Gardens (UK), from which 2,700 were successfully germinated. That same year, about 1,900 seedlings were shipped from Kew to Sri Lanka (at that time called Ceylon), and planted in the Botanical Gardens. In 1877, 22 plants reached Singapore and were planted there. Following this successful transfer, Henry Wickham has been credited with the introducing the rubber tree into South-East Asia. It was from these plants that the first vast plantations sprang up in Malaya, Indonesia and Ceylon (Kochhar 1998: 402-3; Hoi Why Kong 2002: 26).

At the beginning of the twentieth century, European capitalists were sufficiently interested to invest in large plantations in South-East Asia. The expansion of plantations was possible due to the sophisticated organization of plantation enterprises with a very disciplined system of cheap labour and an intensive land use. At that time, the boom in the demand for rubber was mainly due to the automobile industry in the United States. In the 1920s, however, uncertainty about both supply and demand (often driven by changing tyre technologies) meant that the market became extremely volatile (Fig. 4.4). To reduce their dependence on Eastern (Asian) rubber, the United States established plantations in West Africa (namely, Liberia). During World War II, the Japanese invasion of South-East Asia was in part motivated by a desire to close off these

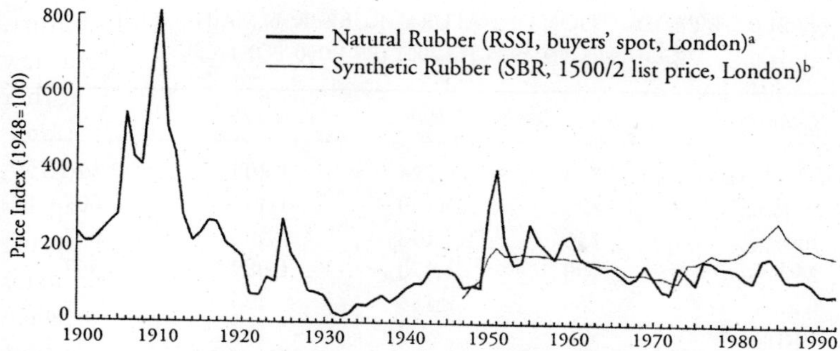


*Source:* NR and SR prices from 1966: International Rubber Study Group (1946-93).

*Notes:* a. From 1948, indexes are in real terms, being deflated by the Manufacturing Unit Value Index in US$ terms of manufactures exported from G-5 countries (World Bank, 1986-93). Before 1948, inflation was minor.

b. Average SBR 1500/2 export prices, New York, up to 1965.

*Source:* Adapted from Colin Barlow et al. (1994).

Fig. 4.4: Index numbers of natural rubber and synthetic rubber, 1900-90.

rubber-producing areas to Western powers. The war provided a stimulus for the Western powers—above and beyond the exploitation of other sources of natural rubber—to develop their synthetic rubber industry. World output of synthetic rubber drew level with that of natural rubber in the early 1960s, and moved ahead thereafter. Despite competition from synthetic rubber, natural rubber continues to play an essential role in industrial production where its specific characteristics such as elasticity, resilience and tackiness are required (Kochhar 1998: 403-4; Zephyr and Musacchio 2002: no pagination; Véron et al. 2004: 22).

Today, the main natural rubber-producing countries are Thailand, Indonesia, India, Malaysia, China and Vietnam (Table 4.3), together accounting for almost 82 per cent of the total world production. Thailand and Indonesia are the leading producer countries with a 32 per cent and 22 per cent total share respectively. In 2001, India was the third largest producer with 632,000 tonnes and a world market share of almost 9 per cent. Thus, rubber production is firmly in countries in South-East and South Asia. Besides Asia, rubber is produced to a small extent in West Africa (Liberia, Ivory Coast) and in South America (mainly Brazil). Over the last twenty-five years, worldwide production has more than doubled: in 2001 it reached a total of 7,210,000 tonnes.

Growth in the consumption of natural rubber during the recent years

TABLE 4.3: PRODUCTION OF NATURAL RUBBER IN MAIN PRODUCING COUNTRIES, 1975 TO 2001 (IN 1,000 TONNES)

| *Country* | *1975* | *1985* | *1995* | *2001* |
|---|---|---|---|---|
| Thailand | 355 | 724 | 1,805 | 2,320 |
| Indonesia | 823 | 1,130 | 1,455 | 1,607 |
| India | 136 | 198 | 500 | 632 |
| Malaysia | 1,459 | 1,470 | 1,089 | 547 |
| China | 69 | 188 | 424 | 464 |
| Vietnam* | 20 | 52 | 155 | 317 |
| Brazil | 19 | 40 | 44 | 88 |
| Total** | 3,315 | 4,400 | 6,070 | 7,210 |

*Source:* *Rubber Statistical Bulletin of the International Rubber Study Group,* in Rubber Board (2003a: 81).

*Note:* *Estimated; **Including those countries not reported separately.

has been driven by a large increase in the demand of emerging developing countries such as China and India, alongside continuing high demand in the US and Japanese markets. In fact, the total demand for rubber (both natural and synthetic)[31] is closely correlated to overall economic growth and industrialization, in particular the growth of the automotive sector (65-70 per cent of the total rubber use is for tyres) (Bruinsma 2003: 122). In 2001, China (with a consumption of 1,215,000 tonnes) was the biggest consumer of natural rubber, followed by the USA (974,000), Japan (729,000), India (679,000), the Republic of Korea (326,000), Malaysia (326,000), Germany (247,000), Brazil (241,000) and France (231,000). The high internal consumption of natural rubber differentiates China and India from other East Asian producer countries. In fact, Thailand, Indonesia, Vietnam and Malaysia have net export shares of 88 per cent, 93 per cent, 92 per cent and 150 per cent[32] respectively, while India and China both had an internal consumption surpassing their own production and were thus net importers.[33]

The future prospects of natural rubber depend to a large extent on the competitive position of natural rubber *vis-à-vis* its synthetic substitute. This position depends on the relative price of synthetic rubber, which will depend on the price of petroleum, its main feedstock. Bruinsma (2003) expects that the share of natural rubber may not grow in the future, due to an increase in its price over the coming years. A projection study suggests that the price of natural rubber will increase until 2010 and then stabilize; the same projection foresees that the share of natural rubber will stabilize at around 37 per cent (Burger and Smit 2001: 22). However, the trend towards a faster growth in the consumption of natural rubber in developing countries (including China and India) compared with industrial ones is likely to continue, leading to a further increase in developing countries' share of world consumption (Bruinsma 2003: 122).

## 4.3 RUBBER CULTIVATION IN KERALA

### 4.3.1 History, Producers and Production

Natural rubber cultivation was established in India in 1902, when some European planters opened an estate at Thattakad, in central Kerala.[34] The previous experience of rubber planting gained by some of the Europeans in Ceylon and Malaya was of enormous benefit to the emergence and subsequent spread of cultivation in Kerala. In 1910, Mundakayam in Kottayam district (Fig. 4.5) became the leading centre of rubber estates in India, and rubber began to flourish (Kershaw 1953: 227). During the 1920s, the development of rubber continued unhindered, and this first period of growth only ended with the Great Depression (in 1929), causing heavy losses to rubber growers for many years due to the drop in price (Fig. 4.4). In 1934, the first international rubber commodity agreement, the International Rubber Regulation Act (IRRA), was set up with the aim of restricting the production and new planting of rubber. The situation in the Indian rubber market changed suddenly with the conquest of various rubber-producing South-East Asian countries by Japan in 1941-2. In order to provide the Allies with sufficient quantities of rubber, many different measures, such as minimum prices, were adopted in 1942 so as to maximize production and to encourage planting of rubber trees (see section 4.4.2). The rubber growers reacted positively; in the years that followed, the increase in acreage under rubber was enormous (Baak 1997: 147-8). According to Lekshmi and George (2003: 224), the statutory price regulation of natural rubber that was initiated in 1942 was a major milestone in the progressive growth of the area under rubber in the state.

There have been two major expansion phases in the history of natural

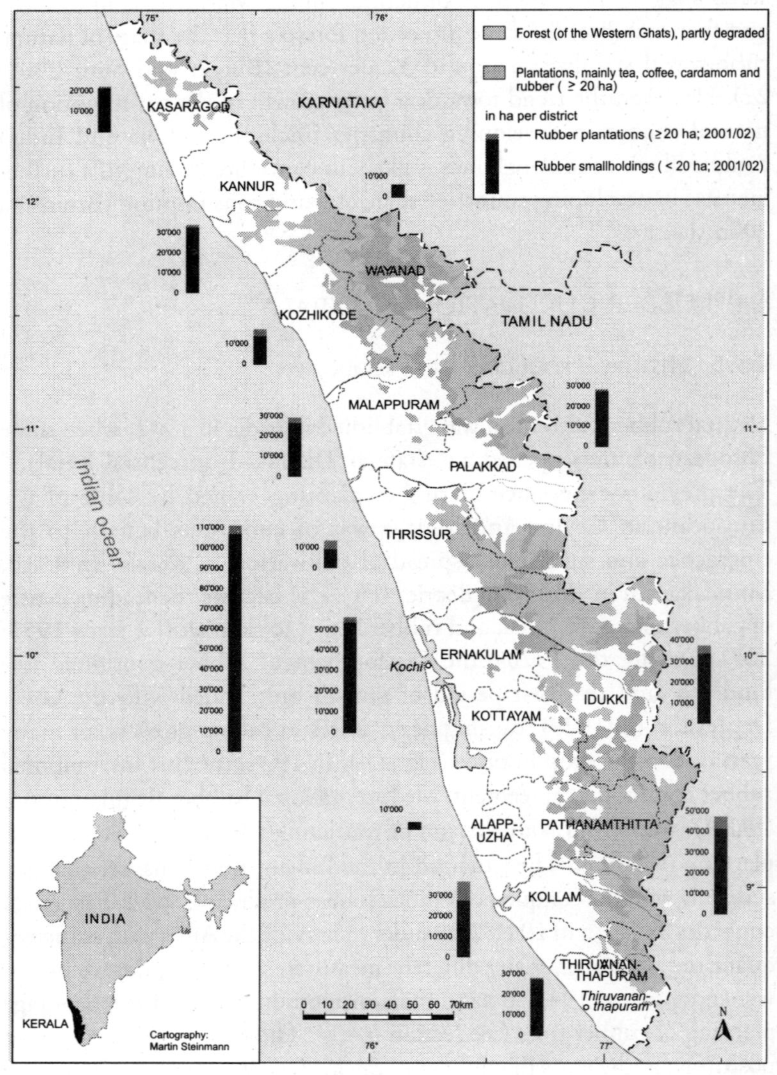


*Source:* Veron et al. (2004, translated from German).

Fig. 4.5: Map of Kerala, with rubber plantations.

rubber in India between the 1950s and today (Table 4.6).[35] The first growth phase started in the 1950s and lasted until the early 1960s. During that time, price regulations and market interventions by different government agencies ensured remunerative prices to growers, with the aim of achieving self-sufficiency in natural rubber production. The Rubber Board and the Rubber Research Institute of India were created in 1947 and 1955, respectively, with the purpose of supporting growers with the necessary institutional framework. In 1957, the Government of India initiated the Replanting Subsidy Scheme which effectively promoted the expansion of the area under natural rubber. During this time, two other policy initiatives, namely the Plantation Labour Act (1951) and the land reform measures (initiated in 1957 by the Government of Kerala) played an important role in the expansion of natural rubber area and a structural change towards smaller holdings. The Plantation Labour Act pushed many farmers to subdivide and fragment their holdings so as to limit the size of their plantation and escape financing the labour welfare scheme for labourers. At the same time, the land reform measures promoted a switch towards natural rubber, since rubber plantations were exempted from land ceiling provisions.[36] These factors, together with the relative profitability of rubber due to government price policy, increased the proportion of the total rubber area under small holdings compared to that under estates (K. Tharian George 1999: 191; Lekshmi and George 2003: 228; Rubber Board 2003a: 5).[37] While the rubber small holdings' share was 44.7 per cent in 1955-6, it rose to 71.6 per cent in 1975-6 (Table 4.4). Currently (in 2000-1), 88 per cent of the total rubber area is on holdings and only 12 per cent on estates. Furthermore, as Table 4.4 shows the size of plantations within the holdings category has greatly diminished over the last fifty years. In 2000-1, 73.3 per cent of all rubber area was ascribed to holdings with 2 ha of rubber or less. The reason small holdings were able to profit from the introduction of rubber was that large investments are not necessary to process the latex collected from the rubber trees. The removal of impurities, coagulation, sheeting and smoking[38] are simple tasks, which the rubber farmers are able to undertake themselves.

The second phase of growth of the area under natural rubber started at the end of the 1970s (Fig. 4.6). This growth was largely triggered by the release of new high-yielding varieties (HYV) in the 1980s, especially the RRII 105 clone.[39] This was a major breakthrough and, together

TABLE 4.4: SIZE SHARE OF RUBBER HOLDINGS AND ESTATES IN INDIA (PER CENTAGE OF TOTAL AREA)

| | *1955-6* | *1975-6* | *1995-6* | *2000-1* |
|---|---|---|---|---|
| Holdings (up to 2 ha) | 23.8% | 39.2% | 71.7% | 73.3% |
| Holdings (> 2 ha up to 4 ha) | 6.6% | 12.4% | 7.5% | 8.0% |
| Holdings (> 4 ha up to 20 ha) | 14.3% | 20.0% | 6.6% | 6.7% |
| *Total of all holdings* | *44.7%* | *71.6%* | *85.8%* | *88.0%* |
| Estates (> 20 ha) | 55.3% | 28.4% | 14.2% | 12.0% |

*Source:* Author's calculations based on Rubber Board figures (2004: 103-4).

with an appropriate package of practices and a coordinated extension service, the country achieved an average yield of 1,576 kg of latex per hectare in 2001-2 (Table 4.5). In 1979, the government launched a credit-linked scheme, which further promoted the new planting and replanting of rubber. The rubber has thus always been an attractive crop for small holdings. However, the shift from other crops (mainly coconut and tapioca) towards natural rubber was also for other important reasons. One reason was the widespread root-wilt disease which has affected the profitability of coconut cultivation since the late 1970s and thus greatly favoured the shift from coconut to rubber (Lekshmi and George 2003: 224). Other reasons are linked to the significant structural transformation in favour of comparatively less labour-intensive, perennial and commercial crops, often at the expense of annual food crops such as rice and tapioca.

Since the 1980s, the annual growth rate of natural rubber has decelerated considerably (Fig. 4.6). This is linked to the fact that the availability of agro-climatically suitable land for further expansion has become limited. Another reason is that while the costs of developing new rubber plantations are steadily increasing (Lekshmi and George 2003: 229) there is a gradual decline in the proportion covered by planting subsidies. However, looking back at the development of natural rubber cultivation since the 1950s, it can be said that its dynamic growth has been due to its relative price stability and profitability compared to other crops, and to fairly good institutional support from the stage of cultivation through to marketing. Nowadays, Kerala holds a near monopoly position in the cultivation and production of natural rubber in India; in 2001-2, the state accounted for 84 per cent of the total area under rubber in India, and for 92 per cent of total national production (Rubber Board

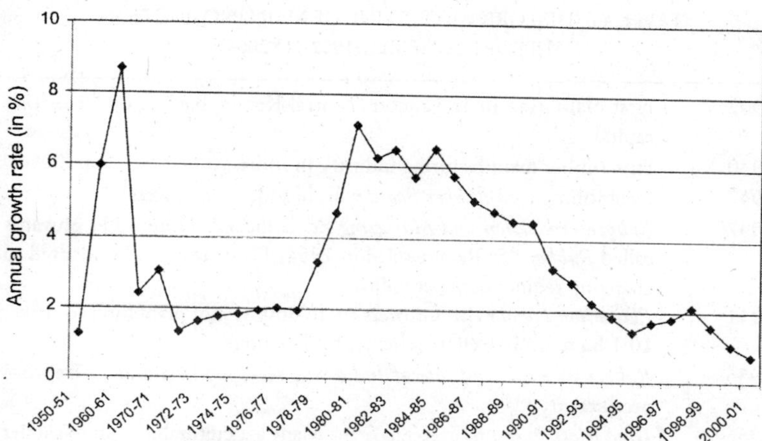


*Source:* Rubber Board (2003a: 2-4).
*Note:* Data on a yearly basis only from 1970-1.

Fig. 4.6: Annual growth rate of area under natural rubber in India, 1950-1 to 2000-1.

TABLE 4.5: TOTAL AREA, PRODUCTION AND PRODUCTIVITY OF NATURAL RUBBER IN INDIA, 1955-6, 1975-6, 1995-6 AND 2001-2

| | *1955-6* | *1975-6* | *1995-6* | *2001-2* |
|---|---|---|---|---|
| Total area (ha) | 86,067 | 235,876 | 524,075 | 566,555 |
| Production (tonnes) | 23,730 | 137,750 | 506,910 | 631,400 |
| Productivity (kg per ha) | 353 | 772 | 1,422 | 1,576 |

*Source:* Rubber Board (2004: 103).

2003a). In that year, the bulk of natural rubber was produced by 997,871 small holdings and, only to a minor extent, by the 307 estates. The average size of the holdings is 0.50 hectare, with 83 per cent of the holdings having 2 hectares or less of rubber (Table 4.4; Rubber Board 2004). Thus, Kerala's natural rubber economy is dominated by small and marginal holdings.[40]

### 4.3.2 Market and Marketing

In India, natural rubber producers mainly transform latex into rubber sheets; sheet rubber's success among the smallholdings is largely due to the fact that this represents a very simple method of processing latex

TABLE 4.6. HISTORICAL OUTLINE OF MAIN PRODUCTION SUPPORT SCHEMES (1902 TO 2004)

| | |
|---|---|
| 1902 | First plantation in Travancore, central Kerala, backed up by British capital. |
| 1930s | First rubber manufacturing industry in India. |
| 1947 | Establishment of *Rubber Board* with headquarter in Kerala. |
| 1947 | *Rubber Production and Marketing Act* launched. From 1954 onwards called *Rubber Act*. Re-amended in 1954, 1960, 1982 to be adapted to changing requirements of industry. |
| 1951 | *Plantation Labour Act*. Growers try to limit size of plantation to max. 10.1 ha to avoid welfare schemes for labourers. |
| 1955 | *Rubber Research Institute of India* was set-up. Main aim to improve productivity. |
| 1956-7 | *Land reform measures* in Kerala State. Many agriculturalists shift to rubber because plantation crops (i.e. rubber) are exempted from land ceiling. |
| 1957 | Launch of *Replantation Subsidy Scheme*. Financial assistance for promoting high-yielding variety planting material. Focus on small growers. Improved from time to time and stopped in 1979. |
| 1960 | First *Rubber Marketing Cooperative Societies* promoted by Rubber Board. |
| 1962 | Set-up of *Plantation Corporation of India* (large plantations under public sector corporations). Development of public sector plantations in northern states. |
| 1962-79 | Three financial schemes: *Upkeep Loan Scheme, Newplanting Loan Scheme, Revised Loan Scheme.* |
| 1970s | Release of new *high-yielding varieties* (HYV), including *RRII 105*, one of the highest yielding in the world. |
| 1979-80 | Launch of *New Planting Subsidy Scheme*. Credit linked scheme for increasing rubber production through cultivation extension (only for small growers). |
| 1980-4 | Launch of *Rubber Plantation Development Scheme* (RPD), phase I. Integrated subsidy scheme for new planting and replanting (for small and big growers). |
| 1985-9 | *RPD Scheme*, phase II. Integrated subsidy scheme for new planting and replanting (limited for growers up to 5 ha in traditional areas). |
| 1986 | Promotion of *Rubber Producer Societies* (RPS) by the Rubber Board. |
| 1990-2 | *RPD Scheme*, phase III (similar as in phase II above). |
| 1991 | Government of India launches *New Economic Policy.* |
| 1993-2001 | *RPD Scheme*, phase IV (similar as in phase III above). |
| 1994 | Government of India becomes member of WTO. |
| since 2003 | *RPD Scheme*, phase V (similar as in phase IV above). |

*Sources:* Author's figures based on Kees Burger et al. (1995), K. Tharian George et al. (1988), George (1999), and Rubber Board (2004: 103).

into a marketable form. Currently, the majority of natural rubber production—69.3 per cent—is marketed in sheet rubber form, 11.5 per cent in the form of latex concentrates, 11.5 per cent in the form of solid block rubber, and the rest, 7.7 per cent, in other forms (Rubber Board 2003a). Thus, the whole Indian natural rubber marketing chain—from the village dealer to the retailer—is concentrated on the marketing of rubber sheets and scrap rubber.

At a glance, the marketing chain is as follows. Smallholders bring the sheets to village dealers, usually their regular customers, who will weigh and grade the rubber sheets according to different RSS categories.[41] Grading is carried out by visual examination, normally by holding rubber sheets against the light which shows up the most obvious defects. The farmer is then paid according to weight and grade. The rubber is stocked—depending on the price and the stocking capacity of the village dealer—until a wholesaler collects it. The wholesale dealer buys at a regional level and bridges the gap between the local trader and the processing or manufacturing unit. Sometimes wholesale dealers also act as agency dealers who will market the product on behalf of producers (de Haan et al. 2003: 190). The agency dealers normally work for large manufacturers, especially in the tyre industry which consumes the bulk of Kerala's sheet rubber production. In order to market rubber, dealers must be licenced by the Rubber Board. In 2001-2, 8,412 dealers were licenced in Kerala; 2,304 of them were active in the Kottayam District (Rubber Board 2003a: 67). Since the 1960s, the Rubber Board has pushed for the development of marketing cooperatives at the district level and producers' societies (RPS) at the village level, as these provide an alternative marketing channel for farmers (section 4.4.1). Today, 285 such societies are active in India, and in 2001-2 they marketed 19 per cent of the total natural rubber production (75). To allow for the production of rubber goods, rubber must first be processed into a form suitable for the manufacturing industry. Subsequently, rubber is transformed into one of the many rubber products. Concerning the marketing of the rubber, it is worth mentioning that there has been widespread smuggling of rubber sheets across the Kerala borders into other south Indian states in recent years. The reason behind this is that sales tax in Kerala is high (12 per cent) compared to Tamil Nadu (8 per cent) and Karnataka (4 per cent), making it lucrative to smuggle sheets into other states to market them. This is expected to lead to heavy losses in tax revenue in Kerala (*The Hindu* 2003).

Natural rubber can be used in a wide variety of ways and the industry is therefore very diversified. Certain low-end products (e.g., Hawaii chappals[42]) can be produced from locally traded, ungraded and scrap rubber; others, such as high-end dipped goods for the international market (e.g., condoms and gloves for medical use), require sophisticated manufacturing companies (Mokhtar 2006: 68-70). In the Indian manufacturing sector, rubber is mainly consumed for vehicle tyres and tubes (44 per cent), cycle tyres and tubes (13.3 per cent), footwear—both as parts for shoes and for Hawaii chappals—(11.2 per cent), belts and hoses for industrial purpose (6.1 per cent), camel backs[43] (6 per cent), dipped goods (3.7 per cent), latex foam (3.7 per cent) and others (10.2 per cent) (Rubber Board 2003a: 58a).

Contrary to expectations, the rubber manufacturing sector in Kerala consumes only around 14 per cent of the state's rubber production, although it tops the list in terms of the number of manufacturing units (Krishnakumar 1997: 233). Most of the larger rubber-consuming companies are located elsewhere in India, especially tyre companies and companies producing high value-added products such as dipped goods. The simpler products are produced locally in small production units using mainly lower quality rubber (de Haan et al. 2003: 190). However, this rubber industry is important for Kerala because of the employment it generates.[44]

## 4.4 INSTITUTIONAL CONTEXT OF RUBBER CULTIVATION IN KERALA

The successful development of the natural rubber sector in India has been made possible by among other things, good institutional support from production to marketing. Since the 1950s, many support mechanisms have been implemented with improving production for achieving self-sufficiency as their main aim. Section 4.4.1 details the main support schemes developed for smallholders[45] and implemented from the 1950s until the present day. Section 4.4.2 gives an overview of the principal price policies in the Indian rubber sector. Section 4.4.3 concludes with an overview of the future challenges for the sector.

### 4.4.1 Production Support Since Independence

The first support scheme for the improvement of natural rubber cultivation, the Replantation Subsidy Scheme, was launched in 1957.

One of the main characteristics of the scheme was that, along with financial assistance, concrete steps were taken to promote the replantation of existing production areas with new planting material, especially with high-yielding varieties (George et al. 1988: 161). With occasional improvements and updates, the scheme remained in force until 1979.[46] Over the twenty-two years that the scheme was in force, replanting was carried out on a total area of 53,605 hectares under 34,822 replanting permits (Rubber Board 2004: 82). In 1979, a comprehensive credit-linked subsidy scheme named the New Planting Subsidy Scheme was launched. The objective of this one-year pilot scheme was to increase production by extending production. The scheme was applicable only to small growers and foresaw a capital subsidy of Rs. 5,000 for an area up to 20.23 hectares (to the equivalent of 50 acres).

This successful pilot scheme was succeeded by an integrated scheme, the Rubber Plantation Development Scheme (RPD), which started in 1980. In this scheme new planting and replanting subsidy schemes were merged and were given equal importance. Since its inauguration, the RPD Scheme has gone through different phases, namely Phase I (1980-4), Phase II (1985-9), Phase III (1990-2), Phase IV (1993-2001), and, from 2003 to the present day, Phase V. Under Phase I to Phase III, participating growers owning up to 5 ha in traditional rubber-growing areas were entitled to receive a planting grant of Rs. 5,000 per ha in annual instalments, spread over 6 to 7 years during the immature phase of the plantations. Additional assistance was granted for the use of poly-bagged plants.[47] Furthermore, as a supplement to the planting grant, participating growers had access to long-term bank loans under NABARD's[48] refinancing scheme, by which the Rubber Board paid 3 per cent interest directly to the bank. Under the RPD Scheme Phase IV, the grant was updated to take into account the increasing costs of production. Up to 1996, it was Rs. 8,000 per ha; in 1997 it was increased to Rs. 18,000 per ha, but limited to 2 ha; and in 2000, the grant was decreased to Rs. 12,000 per ha for areas up to 5 ha. A farmer was required to plant a minimum of 0.10 ha. The polybag plant grant remained in force (although the size of the grant varied); only the subsidy of interest on bank loans was discontinued. All participating growers received free extension support for all stages of planting, maintenance, tapping and processing. The current Phase V of the RPD Scheme was the result of modifications to Phase IV, incorporating various components in addition to replanting and new planting subsidies. The measures under the current RPD scheme are as follows (Rubber Board 2004: 84):

a. Replanting and new planting
b. Measures of productivity enhancement in smallholdings
c. Assistance for beekeeping in rubber plantations as an extra income generation activity
d. Assistance for formation and strengthening of Rubber Producer Societies
e. Quality planting material generation and distribution
f. Farmer's educational programmes
g. Support for maintenance, harvesting and group processing of mature plantations raised under the *tribal development programmes*[49]

Besides implementing these schemes, the Rubber Board and the Rubber Research Institute have—since their creation in the 1950s been actively involved in research, especially the development of new planting varieties and their appropriate package of practices, and the development of processing technologies. These were propagated through an extensive and well-coordinated extension scheme (George 1999: 190). All the schemes mentioned above and the institutional involvement are clear indicators that the government sees raising productivity in the traditional natural rubber areas as one means of achieving self-sufficiency in production.

On the marketing front, Kerala has seen the promotion of two main institutions. From the 1960s onwards, the Rubber Board started promoting Rubber Marketing Cooperative Societies with the aim of encouraging farmers to organize themselves. The Board mainly offers organizational assistance, share participations and working capital loans. Besides rubber marketing, these cooperatives are responsible for the marketing of various plantation inputs such as fertilizers, fungicides, acid to make latex coagulate and tapping aids. There are presently roughly forty functioning rubber marketing cooperatives, operating at the level of either a district or a taluk (Rubber Board 2004). However, it has been established that these societies do not reach a large number of smaller growers, particularly poorer growers in rural areas. The political and bureaucratic controls in these societies have also hindered the promotion of the concept of self-help. Since 1986, the Rubber Board has promoted the formation of small voluntary associations of small growers called

Rubber Producers' Societies (RPS), in order to overcome these problems. According to the Rubber Board, this concept has been widely accepted by the grower community and at present there are 2,100 RPS in the country.[50] RPS function as self-help groups at the village level under the guidance of the Rubber Board. According to the Rubber Board, 'RPS can help the devolution of extension functions, leading to empowerment of the grower community'.[51]

### 4.4.2 Price Policies and Developments in Natural Rubber Prices

Trade policies and control measures in the Indian natural rubber sector since independence can be separated into two distinct phases: a 'protected trade policy' phase (1947-91) and a 'post-1991' phase (1992-today). The major price support mechanisms are highlighted in Table 4.7.

During the first phase, price and trade policies focused on the goal of Indian self-sufficiency in natural rubber production. A turning point in the history of the natural rubber sector came in 1942, when rubber was brought under price control for the first time under the Rubber Control and Production Order. This marked the beginning of governmental price regulation for rubber in India. After the World War II, with the boost of increasing rubber consumption by domestic industry, the government decided to continue with the application of regulatory mechanisms under the Rubber Production and Marketing Act (called the Rubber Act after 1954).

Until 1991, the government used different mechanisms to control the natural rubber market, such as minimum and maximum prices, buffer stocks and control of exports and imports through tariff and non-tariff barriers. The main aim of these policies was to ensure a remunerative price so as to provide incentives to growers to contribute to self-sufficiency. Until 1973-4, fluctuations in prices were mainly intra-year in nature, with great seasonal variations. Since 1974-5, there have been fluctuations (Fig. 4.7) on account of important changes in policy regarding imports, exports and the amount of raw material stocks to be held by manufacturers (George et al. 1988: 160-2). The second '*post-1991*' phase started at the beginning of the 1990s, when the natural rubber sector was integrated into the *New Economic Policy* of the Indian government (including India's entrance into the WTO on 15 April 1994). A first major policy shift occurred when direct imports of rubber by

TABLE 4.7: HISTORICAL OUTLINE OF PRICE SUPPORT (FROM 1947 TO THE PRESENT DAY)

| | |
|---|---|
| 1942 | First *statutory price regulation* launched by the government of India. |
| 1942-6 | Monopoly procurement of NR at fixed prices by the government, until 1946. |
| 1947-91 | Protected trade policy regime. Focus on self-sufficiency in NR production. |
| 1947 onward | Market control through various policy initiatives with notification *of minimum and maximum prices, buffer stocks, exports* and *control of imports of natural rubber through tariff and non-tariff barriers.* |
| 1968 | *Maximum price* removed from price regulation. Prices open to more fluctuations. |
| early 1970s | Major policy shift. *State Trading Corporation of India* (STC) is entrusted with control of external trade of NR (until early 1970s imports directly undertaken by manufacturers). |
| 1986 | Introduction of *Buffer Stock Scheme* (BSS) in response to fluctuating prices. Suspended in 1994. |
| 'post-1991' phase | Priority on value-added exports with competitiveness in cost and quality. |
| 1991 | Government of India launches economic reforms under the *New Economic Policy.* |
| 1991-2 | Dilution in the tariff and non-tariff barriers on NR imports. Imports routed through duty-free channels as an incentive for exports of rubber products. Imports through STC are dropped. |
| since 1992 | As a consequence of policy change: synchronization of NR prices in the international and domestic markets since 1992. |
| 1994 | Government of India becomes member of WTO. |
| 1999 | From Feb. Government bans duty-free import of NR under the advance licencing scheme (ALS). |
| 2001 | 31 March: Removal of *quantitative restrictions* (QRs) on NR imports (existing since 1947). NR can be freely imported by paying the prevailing customs duty. Current rate of basic customs duty for NR is 25 per cent. |
| 2001 | 12 September: *Minimum statutory price* (MSP) for RSS 4 and RSS 5 sheet rubber becomes effective. Declared under a Supreme Court order. |
| 2001 | 10 December: Restriction of NR imports only through the port of Kolkata and Visakhapatam becomes effective. |
| 2001 | 12 December: Mandatory quality standards for both domestically processed and imported NR becomes effective. |

*Source:* Author's figures based on K. Tharian George et al. (2002) and *EPW* Editorial (2001b; 2001a).

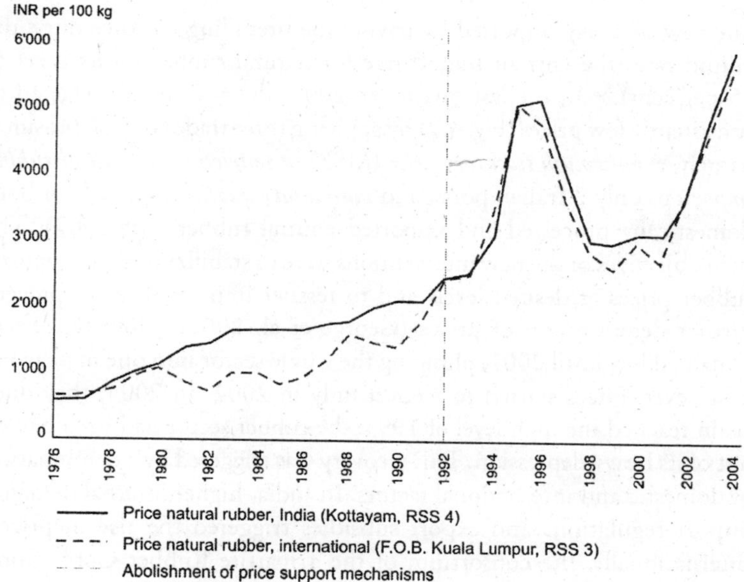


*Source:* Rubber Board (2003a; 2005); Rubber Board website, International price: http://www.rubberboard.org.in/rubberprice.asp?url= international-rubberprice.asp, accessed: January 2006.

*Note:* For price comparison, the international equivalent of the Indian RSS 4 grade (Kottayam price) is the RSS 3 grade (Kuala Lumpur price). Real prices.

Fig. 4.7: Natural rubber price development, Indian and international markets.

manufacturers through duty-free channels were accorded as an incentive for exports of rubber products. Import duty was then reduced as a second stage. A consequence of these changes is that the domestic price of natural rubber has, since 1992, become synchronized with the international price (Fig. 4.7) (George et al. 2002: no pagination).

The convergence of domestic and international prices began to attract more attention when prices started to decline from 1997 onwards. In September 1997, the central and state governments tried to help prices recover through targeted buffer stocks of natural rubber, but this did not lead to the expected result. As a consequence, the government banned imports of natural rubber in February 1999 due to the prevailing uncertainty in the domestic market. On 31 March 2001, quantitative restrictions on natural rubber imports were removed and natural rubber can now be freely imported by paying the prevailing customs duty, the bound rate (the current bound rate for natural rubber sheets is set at 25 per cent).[52] In the last quarter of 2001, the government reacted to consistently low prices (Fig. 4.7) by applying three trade control measures, namely, a *minimum statutory price* (MSP), *a restriction of natural rubber imports* to only 2 Indian ports, and *mandatory quality standards* for both domestically processed and imported natural rubber (Table 4.7). The twin objectives of the new interventions were to stabilize domestic natural rubber prices at desired levels and to restrict imports so as to prevent further deterioration of prices (George et al. 2002). However, prices remained low until 2001, plunging the whole sector into one of its worst crises ever. Prices started to recover only in 2002. In 2004, they once again reached the high level of 1995: the rubber sector had been hauled out of its heavy depression. This recovery was triggered by a combination of domestic and international factors. In India, higher internal demand, import regulations and export subsidies triggered the rise in prices. Internationally, the consortium of the Tripartite Rubber Corporation (including Thailand, Indonesia and Malaysia) played a crucial role in controlling production and thereby ensuring a minimum price for their domestic producers (*EPW* editorial 2004).[53]

The low price phase between 1997 and 2001 (referred to as the 'rubber crisis' in this work) had a drastic impact on many farmers, as rubber was not the only crop affected. In fact, since the mid 1990s, the whole plantation sector had been undergoing a profound crisis, and farmers with other crops than rubber (pepper, coffee, cardamom) were affected by multiple low prices.[54]

### 4.4.3 Future Challenges for the Sector

The Indian natural rubber sector is currently confronted with several new challenges. It has become imperative for the sector to work on producing low-cost and high-quality rubber in order to expand into new export markets, but a rubber sector dominated by smallholdings mostly processing rubber sheets represents a serious constraint. In fact, international markets predominantly require other types of rubber (e.g., TSR—technically specified rubber) rather than rubber sheets. Furthermore, there are few possibilities to lower the cost of production on the farm management side. In the long term only a comprehensive policy package which addresses structural issues, at both farm management and trading levels, will ensure the success of the rubber sector through efforts to enhance the producer's share of the value chain (George and Joseph 2005). The sector must also develop and market the properties of natural rubber, for example, its environment-friendliness[55] and biodegradability (*EPW* editorial 2001a, 2004). Finally, growers should seek additional revenue opportunities, both within the rubber sector (e.g., rubber wood, honey production in rubber plantations) and outside it (Krishnakumar 1997: 233; George 1999: 196-7).

## NOTES

1. This section is mainly based on P.S. George and S. Chattopadhyay (2001: 79-82) and Véron (1999: 76-7).
2. Kerala may have been named after *kera-lam*, which in Sanskrit language means 'the land of coconut', in reference to the abundance of coconut palms in the state (Véron 1999: 72).
3. For a map of Kerala, see Fig. 3.2 or Fig. 4.5.
4. As defined in the Census 1961 (Syamasundaran Nair [no year]), lowland is below 7.6 m above msl, midland is between 7.6 and 76 m above msl and highland is over 76 m above msl. In this classification, however, the category highland encompasses very different agro-ecological zones (including areas of an altitude of more than 2,000 m). Therefore, following Véron (1999: 77), the lower part of the highland (with an altitude of 76 to 400 m above msl) is also called 'mid-highland'. In this study, this lower part of the highland is included in the category 'midland' because topography, climate, soils and cropping patterns are very similar.
5. The division into lowland, midland and highland provides, of course, a very simplified picture which masks the diversity within each of these natural regions. Kerala's Committee on Agro-Climatic Zones and Cropping Patterns (Syamasundaran Nair [no year]), for example, distinguished 13 agro-climatic zones, each with its own relatively distinctive cropping pattern.
6. Rainfall deviation from the normal annual average (sometimes up to 44 per cent) is of significance since it may affect the productivity of perennial crops such as coconut, rubber and pepper in the long run, besides also affecting the production and productivity of annual crops. Between 1999 and 2002, the total deviation from normal was considerable, ranging between -6 per cent to 21 per cent (Kerala State Planning Board 2002: 38).
7. These are Thiruvananthapuram (Trivandrum), Kollam (Quilon), Pathanamthitta, Alappuzha (Alleppey), Kottayam, Idukki, Ernakulam,

Thrissur (Trichur), Palakkad (Palghat), Malappuram, Kozhikode (Calicut), Wayanad (Wynad), Kannur (Cannanore), Kasaragod. The old (English) names, some of which are still widely used, are in parentheses.

8. These are the cities of Thiruvananthapuram and Quilon in south Kerala, Kochi (Cochin) in central Kerala, and Thrissur and Kozhikode (Calicut) in north Kerala. They have between 317,474 (Thrissur) and 744,739 (Thiruvananthapuram) inhabitants.
9. Based on the summary in Heidi Stutz (1999: 78-81).
10. Nowadays, however, formal education has become an increasingly important factor in the choice of profession and the caste system has become much more flexible.
11. Government of India, Directorate of Census Operations, Kerala, http://www.censuskerala.org/state_profile.html. Accessed on 02 January 2006.
12. In 1991, the national average of Christians was 2.3 per cent (Harriss-White 2003: 152).
13. According to the following sources—literacy rate: Census of India 2001 (Government of India 2001: 13-14); infant mortality rate and life expectancy Kerala: Kerala State Planning Board (2003: 28); infant mortality and life expectancy India: Government of India, National Commission on Population, Facts: http://populationcommission.nic.in/facts1.htm, accessed: 24 January 06.
14. According to the National Account Statistics, Government of India, New Delhi: http://macroscan.net/dat/jul02/pdf/Data/N80-71A.pdf, accessed on 24 January 06. There is one important remark to be made when interpreting the growth performance of Kerala: taking the domestic product to represent growth performance is seriously limitating, because of the significant size of remittances Kerala receives from its emigrant workers. It has been estimated that remittance income was around 23 per cent of NSDP during the second half of the 1990s (Kannan 2005: 550).
15. According to Government of Kerala, Department of Public Relations, State Economy: http://www.prd.kerala.gov.in/prd2/eco/state1.htm, accessed: 24 January 06.
16. The primary sector includes agriculture, forestry, fishing, as well as mining and quarrying. Agriculture, forestry and fishing account for 25.1 per cent, mining and quarrying for 0.2 per cent of the total.
17. George and Chattopadhyay (2001: 89-90) also mention the conversion of rice land for non-agricultural uses, such as for the construction of buildings and courtyards, roads and railways, canals and storage tanks. Although the law prohibits paddy land conversion, the rules are rarely enforced.
18. In Kerala, hired agricultural workers traditionally carry out the bulk of the work in the fields. Landowners' participation is often limited to supervising

wage labourers. Most landowners avoid manual labour, which is often associated with social inferiority.

19. The government's differential tax structure promoted market-oriented cultivation especially aimed at export markets. The commercial crop sector in Kerala responded positively to this stimulus by developing a lucrative export market (Panikar et al. 1978 in George and Chattopadhyay 2001: 92).
20. As an example, George and Joseph mention that 'it is possible that the timing and announcement of imports are used to create uncertainties in the primary markets to reap maximum benefits at the tail-enders' level' (2005: 2684).
21. The highly fluctuating prices of commercial crops, together with other uncertainties faced by the agricultural sector with the India's entrance into the WTO, forced the government of Kerala to establish the Commission on WTO Concerns in Agriculture which recently published its recommendations (Government of Kerala 2003).
22. This section is based on Kochhar (1998: 404-10) and the Rubber Board website: http://www.rubberboard.org.in/ManageCultivation.asp?Id=2, accessed: Jan. 2006, and Rubber Board (2004: 3).
23. Natural rubber, however, has been found in the latex of over 200 species of plants belonging to 79 different families (Rubber Board 2004: 3).
24. Though *Hevea* endures periodic flooding, it produces high yields regularly only on well-drained soil (Rehm and Espig 1991: 365).
25. In 2001-2, in India, average yield was 1,576 kg per ha (Rubber Board 2004: 103).
26. Wickham trees are the progenies of the trees imported by Henry Wickham to Asia in 1876.
27. Rubber collectors of the Amazon Basin originally gathered latex by cutting down the wild trees of *Hevea*, but later tapped them by making haphazard wounds with the help of crude tools. In contrast to modern-day tapping techniques, early systems were extremely wasteful and inefficient (Kochhar 1998). Nowadays, tapping robots are in development, but their use is still very expensive (Trueb 1996).
28. Latex from *Hevea* is a white or slightly yellowish opaque liquid with the following composition: water (55-65 per cent), rubber (30-40 per cent), proteins (2-2.5 per cent), resins (1-2 per cent), sugars (1-1.5 per cent) and ash (0.7-0.9 per cent).
29. Latex can be processed into sheet rubber (RSS), preserved field latex (and latex concentrate), block rubber, and crepe rubber. Scrap (field coagulum) can be processed into block or crepe rubber. In India, around 75 per cent of the natural rubber currently produced is marketed in the form of rubber

sheets (Fig. 4.3), as it is the oldest and the simplest method of processing latex into a marketable form (section 4.3.2).

30. Vulcanization refers to the reaction when rubber is combined with sulphur; this reaction converts the plastic and viscous nature of raw rubber into elastic. Vulcanized rubber has very high tensile strength and comparatively low elongation. Its hardness and abrasion resistance are higher than those of raw rubber. Because of the unique combination of these chemical and physical properties, natural rubber finds an application in the manufacture of a variety of products (Rubber Board website, Properties and Uses of Natural Rubber: http://www.rubberboard.org.in/ManageCultivation.asp?Id=192. Accessed on 5 January 2006).
31. The share of natural rubber in total rubber consumption is influenced by prices, technology and the mix of final rubber products. This latter factor can be particularly important; in fact, the increase in the share of radial tyres and those of heavy trucks (which require a higher component of natural rubber) in total tyre production favours the use of natural rubber. These days, natural rubber accounts for about 40 per cent of the aggregate rubber consumption (Bruinsma 2003: 122).
32. Malaysia uses domestic production and imports to manufacture rubber products for export.
33. Consumption-related data from the Rubber Statistical Bulletin of the International Rubber Study Group, in Rubber Board (2004: 110). Export shares are calculated based on export data from the Rubber Board (2004:111).
34. However, at that time, other rubber trees such as *Ficus elastica* were already indigenous to Indian forests. *Ficus elastica* was tapped on a large scale in Assam (north-east India) prior to the introduction of *Hevea brasiliensis* (George et al. 1988: 158).
35 The data in Fig. 4.6 refers to the whole of India. However, already in 1955-6 Kerala state accounted for about 94 per cent of the total area under the crop (Lekshmi and George 2003: 232, 10). Therefore, the changes, in the annual growth rate basically reflect changes in Kerala.
36. For the sake of completeness, it is important to mention that between 1930 and 1970, a major migratory movement of farmers took place to the regions of Malabar (north Kerala) and the hilly regions of Travancore (particularly Idukki district). These farmers opened up new agricultural land that was previously under forest. Though migration was also pushed politically under a 'Grow More Food' campaign (1956), the cropping pattern shifted gradually from seasonal subsistence food crops to perennial commercial crops including natural rubber (George and Chattopadhyay 2001: 92-5). See also Geiser (2001).
37. The Rubber Board follows the following categorization: holdings: 2 ha and

below, above 2 ha and up to 4 ha (included), above 4 ha and up to 20 ha (included); estates: above 20 ha.

38. The lack of a special smokehouse, used by large estates to dry the rubber sheets, can in fact be compensated for by 'sun drying' and 'kitchen smoking'. However, these methods result in a rubber of inferior quality compared to that produced on the large estates in particular, which have the necessary facilities essential for producing high quality rubber. However, the deficiency of not having the best quality rubber has often been overcome by a steady and favourable price offered to Indian rubber producers, at least until the 1990s.
39. RRII 105 is the most popular variety planted today; in a study by Rajasekharan and Veeraputhran (2002: 5), it was shown that between 69 per cent (in southern Kerala) and 96 per cent (in northern Kerala) of the farmers had planted RRII 105 exclusively. The potential and yield realized by RRII 105 is rated as one of the highest in the world (George 1999).
40. The evolution towards the dominance of smallholdings is in line with the changes observed in other major natural rubber-producing countries. The estimated smallholder shares of the rubber sector in Thailand, Indonesia and Malaysia are 95 per cent, 84 per cent and 85 per cent respectively (George 1999: 197, Fig. 14).
41. RSS 1 is the best quality, RSS 5 the worst, while RSS 4 is by far the most common grade of sheet rubber in terms of the volume of domestic production.
42. Cheap and common footwear in many developing countries, made from rubber and plastic.
43. Camel backs are used for retreading tyres.
44. According to George et al. (1988: 158), 'there is no disagreement on the strategic importance of NR [natural rubber] as an industrial raw material since it is estimated that around 35,000 different products ranging from aero-tyres to rubber bands can be manufactured from it'.
45. The schemes related to the development of large estates are not detailed in this section since the study is based on the analysis of the institutional support for small rubber holdings.
46. During the same period (1962-9), three other schemes—the *Upkeep Loan Scheme*, *New Planting Loan Scheme* and the *Revised Loan Scheme*—were implemented; however these involved relatively small financial outlays. These schemes promoted the scientific upkeep of immature fields in small holdings, the expansion of small and marginal holdings into viable units, and more extensive expansion of rubber cultivation in the holding sector [as compared to the plantation sector: note from the author] (Rubber Board 2004: 82).
47. A poly-bagged plant is a rubber seedling grown in a plastic bag. Its main

advantage is that it is already being at an advanced stage of growth when it is planted in the rubber field.

48. Indian National Bank for Agriculture and Rural Development.
49. On tribal development programmes, see section 7.1.2.
50. Presently, 53 RPS are labelled as Model Rubber Producers' Societies (Model RPS). These RPS are supported by the Rubber Board, financially as well as technically, to transform them into model ones to act as demonstration centres for other RPS. Support is given for capacity-building and to develop infrastructural facilities for scientific community processing.
51. Rubber Board website, http://rubberboard.org.in/ManageScheme.asp?Id=33, Rubber Producers' Societies, accessed on January 2006.
52. The *bound rate* defines the maximum import duty (i.e. tariff barrier) for a product. For the different bound rates of rubber products, see George et al. (2003: 19). It is important to mention at this point that the Agreements on Agriculture of the WTO define natural rubber as an industrial raw material and not as an agricultural crop. Thus, it is not possible to impose higher duties on natural rubber. According to Toms Joseph and K.Tharian George (2002: 64), 'the exclusion of NR from the list of agricultural products is inconsistent, illogical and unjustifiable.'
53. Thailand, Indonesia and Malaysia account for 80 per cent of the total world natural rubber output. In 2001, the *Tripartite Rubber Corporation* decided on a production cutback of 150,000 tonnes in 2002 and 200,000 tonnes in 2003, and an export cut by 10 per cent from January 2002 onwards (*EPW* editorial 2001b).
54. See Fig. 6.2 for the price development of relevant cash crops in Kerala (from 1990 to the present).
55. See Jacob (2005) on the opportunities for the Indian plantation sector (incl. natural rubber) under the Kyoto Protocol.

CHAPTER 5

# Rubber Holdings in Thalanadu

## 5.1 THALANADU PANCHAYAT

### 5.1.1 Location, Agro-Ecology and Cropping Pattern History

Thalanadu Panchayat[1] is located in the most easterly part of Kottayam district. It can be reached via the small town of Teekoy, which lies on the main road between the town of Eratuppetta and the village of Vagamon in the higher elevations of Idukki district. The distance to Eratuppetta—the main commercial and service centre in the region—is around 7 km, while Vagamon is 10 km away. Thalanadu Panchayat is bordered to the east by Elappara Panchayat, to the west and north by Moonilavu Panchayat, and to the south by Teekoy Panchayat. It covers an area of 32.24 sq km and has a population of nearly 8,200 people living in around 1,550 households.[2] The majority of the population is Christian (*c.* 65 per cent), followed by Hindus (*c.* 25-30 per cent) and Muslims (*c.* 5-10 per cent). This is in line with the average for the central districts of Kerala (see section 4.1.1). Until 2003, the Panchayat was divided into eight wards (the smallest political-administrative unit of a Panchayat), but due to a recent revision it now comprises ten wards (Fig. 5.1).[3]

Thalanadu lies in the lower highlands of central Kerala. The topography is characterized by narrow valleys and hills with steep gradients and slopes. The main centre of Thalanadu lies on a hill. Smaller streams flow into the local Meenachil River, which forms the border of the Panchayat. Most of the hilly areas have laterite soils (Syamasundaran Nair, no year). The soil quality, however, varies consistently at the micro-level, from deeper and more fertile in the lower areas to shallow and rocky in hilly and steep areas.[4] The annual rainfall is high at around 3,200 mm. Most rain—around 2,000 mm—falls between June and September (during the south-west monsoon); about 700 mm falls between October and November (north-east monsoon) and 500 mm falls from December to May (dry season). The mean temperature is 28 °C without much seasonal variation. The minimum temperature at night—in December and January—is 22 °C; the maximum daytime temperature—in March and April—is 33 °C (Centre for Earth Science Studies 1984).

The vegetation in Thalanadu consists almost entirely of intensive, market-oriented agriculture, and almost all cultivable land is under agriculture. The cropping pattern is comprised of different types of perennial, annual and seasonal crops (Table 5.1). However, 75 per cent of the total agricultural land is under natural rubber plantation, and rubber dominates the landscape. Nowadays, agriculture—especially rubber—forms the backbone of the local economy in Thalanadu. However, economic activities in the tertiary sector (e.g., salaried employment, trade, transportation) have become important (see chapter 6). Yet, trade and transportation are still strongly linked to agriculture.

Land use in Thalanadu has changed significantly over the past hundred years, as is the case in most parts of Kerala. The first settlers arrived from the surrounding areas towards the end of the nineteenth century to buy land.[5] At the time, grasslands and forest areas dominated the landscape. It is presumed that the first settlers practised slash-and-burn cultivation with different staple crops (paddy, tapioca). Later, parts of the forest were cleared and the land came under permanent cultivation. The first perennial crops were arecanut, coconut and cashew, as well as tamarind,

TABLE 5.1: AGRICULTURAL PATTERN OF THALANADU PANCHAYAT

| | | | |
|---|---|---|---|
| Total area | 3,224 ha | Cardamom | 28 ha |
| Total agricultural land | 3,067 ha | Mango & Jackfruit | 22 ha |
| of which under single crop | 2,791 ha | Cocoa | 20 ha |
| of which under mixed crop | 276 ha | Teak | 18 ha |
| *Crops* | | Cashew | 15 ha |
| Rubber | 2,312 ha | Vegetables | 10 ha |
| Coconut | 148 ha | Vanilla | 8 ha |
| Pepper | 112 ha | Nutmeg (*Djadi*) | 7 ha |
| Banana (incl. plantain) | 104 ha | Turmeric | 6 ha |
| Tapioca | 40 ha | Ginger | 6 ha |
| Cloves (*Gramboo*) | 28 ha | Arecanut (*Kamughé*) | 5 ha |
| Coffee | 28 ha | Tuber crops | 4 ha |

*Source:* Agricultural office, Thalanadu Panchayat, 17 January 2004.

jackfruit, mango and other tree crops. Once the trade in agricultural commodities picked up, cash crops such as pepper and coffee were introduced. Some interviewees recall that dry land paddy was once found on some terraces in the Panchayat. However, rice cultivation never became common in Thalanadu, and since the 1960s paddy has completely disappeared from the area. In some places at higher altitudes, small tea plantations were established.

In the 1920s, the British started the Teekoy Rubber Estates and introduced rubber cultivation to the Panchayat. The estates were instrumental in the development of rubber plantation amongst the surrounding smallholdings. In fact, some local farmers considered the professional management of the Teekoy Rubber Estates to be a promising sign that rubber was an interesting crop for the future.[6] However, general interest in rubber at that time was still low because there was no market for latex and the local varieties planted at the time gave a very low yield.[7] Interest in rubber cultivation grew during the 1950s and 1960s, at a time when a severe outbreak of disease destroyed the local coconut plants (this subsequently occurred in most districts of Kerala. The first high-yielding varieties (HYV) of rubber and subsidy schemes were introduced by the Rubber Board. During the 1970s and 1980s, most farmers began to plant rubber.

### 5.1.2 Geographical and Social Characteristics of Choovoor and Maravikkallu Wards

The wards of Choovoor and Maravikkallu were selected for this study because of their differing geographical and social characteristics. Choovoor ward lies between the higher areas of the Panchayat (near the main areas of settlement) and the settlement of Moonilavu at lower altitude. The ward covers a vast area and most people live and cultivate land on the steep hillsides. The elevation ranges from 400 to 900 m above msl. The land is quite hilly, steep and rocky, and rubber is planted up to a maximum of 800 m above msl. Until 2004, road access was poor, but in 2005 the construction of a new road between Choovoor and Moonilavu was entering its final stages.[8] However, many houses can still only be reached by steep footpaths.

Maravikkallu ward lies at an elevation of approximately 200 to 500 m above msl and is less hilly than Choovoor. It is adjacent to the small

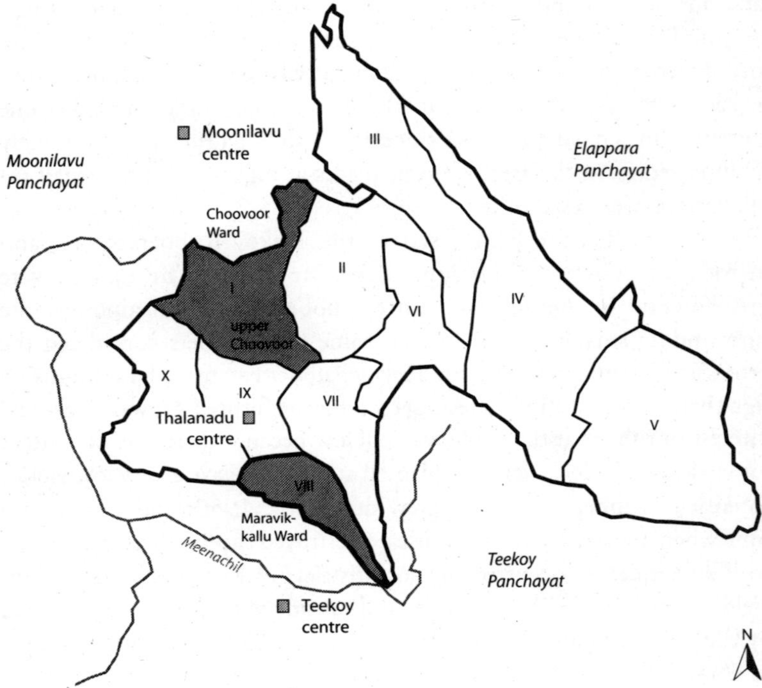


*Source:* Author's figure; not to scale.

Fig. 5.1: Map of Thalanadu Panchayat, with Choovoor ward (I) and Maravikkallu ward (VIII), and main centres.

commercial town of Teekoy, which lies on an important district road. Road infrastructure was developed earlier in this ward, and most houses have long been accessible by road.

The main reason for choosing Thalanadu Panchayat for this study was that most of it lies in a marginal growing area for natural rubber. As a rule, rubber grows best up to altitudes of 300 m above msl, while some of the wards in the Panchayat reach 900 m above msl. Although rubber can still grow at this altitude, there is an impact on the average yield of the trees; thus farmers at higher elevations can expect lower yields, and consequently lower earnings. One of the main differences between Choovoor and Maravikkallu thus lies in their elevation. However, as will be seen in the following section, Choovoor and Maravikkallu also have quite different social realities.

### 5.1.3 Government Institutions and Private Service Sector in Thalanadu Panchayat

Thalanadu Panchayat has a variety of government and private service institutions. Government institutions include the Thalanadu Panchayat office (the local government office), the agricultural office, the post office, the cooperative bank, the cooperative society (delivering subsidized manure, for example), government schools (lower primary and upper grade schools)[9] and a government health centre. Private banks, money-lenders, as well as Ayurveda doctors, are examples of private service institutions in the Panchayat. Private trading institutions include traders and wholesalers of agricultural commodities, retail shops for general consumer goods (including greengrocers), petty trade stalls, some restaurants (*hotels* in the local language), tea shops and toddy shops (the local bar). Other institutions include the offices of the local political parties, churches of the different Christian communities, a temple, and some community centres (e.g., a stitching centre of the local Hindu community). Most of these institutions are located in Thalanadu Centre, the main market area of the Panchayat. However, it is important to note that people from the Panchayat are also oriented towards the two neighbouring centres of Teekoy (in the south) and Moonilavu (in the north) (Fig. 5.1). There is further analysis of this fact in section 7.

## 5.2 RUBBER HOLDERS IN CHOOVOOR AND MARAVIKKALLU

This section describes the results of a short and focused survey (a *household listing*), carried out amongst all the households in the two wards. The first part describes findings relating to *all* households, while the second part relates only to households which actually planted rubber in the period from 1995-6 to 2003-4; these have been called *rubber holdings* for the purpose of this study.

### 5.2.1 Characteristics of Households

*Household size*

All 299 households were surveyed in the two wards: 164 in Choovoor and 135 in Maravikkallu. On average, there were 4.5 persons living in

Fig. 5.2: Marginal rubber holder in the study location (1 May 2004).

Fig. 5.3: A rubber plantation in the study location (24 April 2005).

each household. In Choovoor, the average is slightly lower than in Maravikkallu, with 4.26 and 4.81 persons per household respectively. Table 5.2 shows that Choovoor has more households with 1-2 persons (11 per cent) than Maravikkallu (4.4 per cent); in total, 61 per cent of

TABLE 5.2: HOUSEHOLD SIZE, BY WARD ($n$ = 299)

| *Persons per household* | *Total* | | *Choovoor* | | *Maravikkallu* | |
|---|---|---|---|---|---|---|
| | *Total* | *%* | *Total* | *%* | *Total* | *%* |
| Total | 299 | 100.0 | 164 | 100.0 | 135 | 100.0 |
| 1-2 | 24 | 8.3 | 18 | 11.0 | 6 | 4.4 |
| 3-4 | 141 | 47.2 | 82 | 50.0 | 76 | 43.7 |
| 5-6 | 107 | 35.8 | 55 | 33.6 | 52 | 38.5 |
| 7-8 | 17 | 5.7 | 4 | 2.4 | 13 | 9.6 |
| 9-10 | 10 | 3.3 | 5 | 3.0 | 5 | 3.7 |

*Source:* Author's survey.

the households in Choovoor have 4 or less people, while they represent 48.1 per cent of the total households in Maravikkallu. Nevertheless, Maravikkallu has larger households: 9.6 per cent of its households have 7-8 persons while this accounts for only 2.4 per cent of all households in Choovoor. There are two hypotheses for this: the first is that Maravikkallu attracts more families with numerous children since its infrastructure (roads, schools) is better developed; the second is that households in Maravikkallu are wealthier and can therefore support more children. Both remain hypotheses, however, since the number of children was not surveyed.

### *Religion and caste*

The survey confirms that the Panchayat is situated in a district with a high percentage of Christians. Almost 69 per cent of the households of the two wards belong to the Christian church, 20 per cent are Hindus and the remaining 11 per cent Muslims (Table 5.3). There are, however, differences between Choovoor and Maravikkallu. Choovoor has one neighbourhood where many Muslim families live. Thus, 19.5 per cent of its total households are Muslim, 23.2 per cent Hindus and the majority, 57.3 per cent, Christians. In Maravikkallu, no Muslim households were identified: 83.0 per cent of all households are part of the Christian community, and the rest, 16.3 per cent, are Hindu (with one answer not given). This shows that Hindus are equally distributed over the two wards.

TABLE 5.3: RELIGION AND CASTE AFFILIATION, BY WARD ($n$=299)

| *Religion/Caste* | *Total* | | *Choovoor* | | *Maravikkallu* | |
|---|---|---|---|---|---|---|
| | *Total* | *%* | *Total* | *%* | *Total* | *%* |
| Christian (all) | 206 | 68.9 | 94 | 57.3 | 112 | 83 |
| SC/ST (CSI) | 72 | 24.1 | 72 | 43.9 | 0 | 0 |
| Roman Catholic | 134 | 44.8 | 22 | 13.4 | 112 | 83 |
| Hindu (all) | 60 | 20.1 | 38 | 23.2 | 22 | 16.3 |
| Ezhava | 36 | 12 | 21 | 12.8 | 15 | 11.1 |
| Nair | 3 | 1 | 3 | 1.8 | 0 | 0 |
| SC/ST Hindu | 5 | 1.7 | 0 | 0 | 5 | 3.7 |
| unknown | 16 | 5.4 | 14 | 8.5 | 2 | 1.5 |
| Muslim (all) | 32 | 10.7 | 32 | 19.5 | 0 | 0 |

*Source:* Author's survey.

*Note:* In Maravikkallu, one answer was not given. SC/ST stands for Scheduled Castes and Scheduled Tribes.

To understand better the implications of religious affiliation on income generation, it is necessary to show how many households belong to the Scheduled Castes or Scheduled Tribes (SC/ST). This is important since communities belonging to the SC/ST are eligible for state support under a system of educational and job reservations. Table 5.3 differentiates between households on the basis of their caste affiliation. It is important to note the difference between Choovoor and Maravikkallu in terms of Christians: 43.9 per cent of all households in Choovoor belong to the community of the Church of South India (commonly abbreviated CSI), while there are no families from this community in Maravikkallu.

Within the Hindu community, slightly more than half of the households belong to the Ezhava caste, the caste of coconut and toddy tappers; only five households are from Hindu SC/ST communities. There are no SC/ST divisions in the Muslim community.

### *Land ownership*

Land ownership is defined in this study as the total land belonging to all members of a household. Table 5.4 shows the results of the survey using the Indian government's official land categorization of holdings which is defined as follows: marginal holdings (< 1 ha), small holdings (1-2 ha), semi-medium holdings (2-4 ha) and medium holdings (4-10 ha).[10] The

TABLE 5.4: TOTAL LAND OWNERSHIP OF HOUSEHOLDS, BY WARD (*n*=299)

| *Land size* | *Total* | | *Choovoor* | | *Maravikkallu* | |
|---|---|---|---|---|---|---|
| | *Total* | *%* | *Total* | *%* | *Total* | *%* |
| Marginal holding (> 0-ha) | 201 | 67.2 | 128 | 78.1 | 73 | 54.0 |
| Small holding (> 1-2 ha) | 58 | 19.4 | 22 | 13.4 | 36 | 26.7 |
| Semi-medium holding (> 2-4 ha) | 32 | 10.7 | 11 | 6.7 | 21 | 15.6 |
| Medium holding (> 4-10ha) | 6 | 2.0 | 2 | 1.2 | 4 | 3.0 |
| No land | 2 | 0.7 | 1 | 0.6 | 1 | 0.7 |

*Source:* Author's survey.

majority (67.2 per cent of the total number of households) have a total landholding of one hectare or less. In Choovoor, this percentage is higher (78.1 per cent) than in Maravikkallu (54 per cent). Overall, the situation in the two wards reflects the typical structure within Kerala where many small houses have a homestead garden and there are only a few houses on larger properties. Finally, the survey revealed no land property bigger than 20 ha (qualified as an 'estate') in either of the two wards.

### *Main income-generating activity*

Household representatives were asked to assess which sector of activity provided the main source of household income.[11] Table 5.5 gives the details of the results. Before going into further analysis, it is important to acknowledge that, for many households, secondary incomes can also play an important role in their total income.

TABLE 5.5: MAIN INCOME OF HOUSEHOLDS, BY WARD (*n* = 299)

| *Main income* | *Total* | | *Choovoor* | | *Maravikkallu* | |
|---|---|---|---|---|---|---|
| | *Total* | *%* | *Total* | *%* | *Total* | *%* |
| Farming/agriculture | 148 | 49.5 | 62 | 37.8 | 86 | 63.7 |
| Agricultural labour | 82 | 27.4 | 54 | 32.9 | 28 | 20.7 |
| Non-farm labour | 21 | 7.0 | 13 | 7.9 | 8 | 5.9 |
| Regular salaried job | 30 | 10.0 | 24 | 14.6 | 6 | 4.4 |
| Trade or business | 16 | 5.4 | 9 | 5.5 | 7 | 5.2 |
| Others | 2 | 0.7 | 2 | 1.2 | – | – |

*Source:* Author's survey.

The results give an agrarian picture: almost half (49.5 per cent) of all the households in the two wards derive their main income from 'farming/ agriculture'. The second-most important source of income is work as agricultural labour (27.4 per cent). Only 23.4 per cent said that the main income was derived from activities not directly linked to agriculture, namely, 'non-farm labour', 'regular salaried jobs', and 'trade or business'.

The analysis of the main sources of income per ward suggests that, on average, the livelihood of households in Maravikkallu is more dependent on 'farming/agriculture' (63.7 per cent) than on agricultural labour (20.7 per cent). This is different in Choovoor, where only 37.8 per cent of all households have their main income from 'farming/agriculture', while many other households (32.9 per cent) depend on agricultural labour. Income from regular salaried jobs is also important in Choovoor. A first rough conclusion is that income sources are more diverse across all households in Choovoor than in Maravikkallu.

Table 5.6 shows the main income of households in both wards according to land size. As expected, the greater the amount of land a household owns, the more dependent it is on 'farming/agriculture' for its income. 'Farming/agriculture' consistently constitutes the main source of income for over 80 per cent of all households and for every category of landownership over one hectare. In the smallest category (0-0.5 ha), half of the households (50.6 per cent) depend on agricultural labour, whereas only 16.9 per cent of these households derive their main income from 'farming/agriculture'. Non-farm labour and regular salaried jobs

TABLE 5.6: MAIN INCOME OF HOUSEHOLDS PER LAND OWNERSHIP, IN PER CENT ($n = 299$)

| *Main income* | *0-0.5 ha* | *> 0.5-1 ha* | *> 1-2 ha* | *> 2-4 ha* | *> 4 ha* |
|---|---|---|---|---|---|
| Farming/agriculture | 16.9 | 78.7 | 84.5 | 93.8 | 83.3 |
| Agricultural labour | 50.6 | 6.4 | – | 3.1 | – |
| Non-farm labour | 11.7 | 4.3 | – | – | – |
| Regular salaried job | 11.7 | 10.6 | 10.3 | 3.1 | – |
| Trade or business | 7.8 | – | 5.2 | - | 16.7 |
| Others | 1.3 | – | – | – | – |
| Total | 100.0 | 100.0 | 100.0 | 100.0 | 100.0 |

*Source:* Author's survey.

have the same importance as income from 'farming/agriculture' (each with 11.7 per cent). It can thus be concluded that households with less than 0.5 ha of land can rarely derive their main income from 'farming/ agriculture'.

As a final remark, the importance of the categories 'regular salaried job' and 'trade or business' seems be somewhat dependent on the size of the holding: 16.7 per cent of the largest holdings stated that their main income came from trade or business. This is surprising since one would expect bigger holdings to be more concentrated on their farm and therefore have less time for non-farm activities. Conversely, regular salaried jobs are relatively more important for holdings with less than 2 ha.

### 5.2.2 Characteristics of Natural Rubber Holdings

From this point on, the study will focus on rubber holdings, i.e. households who either currently plant or have previously planted natural rubber as an income activity. The survey of all 299 households in the two wards shows the following results (Table 5.7): on average, over the two wards, 216 households (72.2 per cent) had natural rubber on their land, while 83 households (27.8 per cent) did not cultivate rubber. The proportion of rubber holdings is smaller in Choovoor (64.6 per cent) than in Maravikkallu (81.5 per cent).

Table 5.8 shows the main characteristics of rubber holdings in Choovoor and Maravikkallu. Analysis of the total landownership of these holdings—including both plantations[12] of natural rubber and other cultivation—shows that they are smaller in Choovoor than in Maravikkallu.[13] In Choovoor, 47 per cent of all holdings have 0.5 ha or less, and 88 per cent have 2 ha or less; in Maravikkallu, only 22 per cent

TABLE 5.7: RUBBER HOLDINGS, BY WARD ($n$ = 299)

| | *Total* | | *Choovoor* | | *Maravikkallu* | |
|---|---|---|---|---|---|---|
| | *Total* | *%* | *Total* | *%* | *Total* | *%* |
| Rubber holdings | 216 | 72.2 | 106 | 64.6 | 110 | 81.5 |
| Households without rubber | 83 | 27.8 | 58 | 35.4 | 25 | 18.5 |

*Source:* Author's survey.

TABLE 5.8: CHARACTERISTICS OF RUBBER HOLDINGS IN CHOOVOOR AND MARAVIKKALLU WARDS ($n$ = 216)

| *Category* | *Total* | *Choovoor ward* | *Maravikkallu ward* | *Survey RRII (George, 1999)* |
|---|---|---|---|---|
| Total rubber holdings | 216 | 106 | 110 | 2,575 |
| Size category | | | | |
| Marginal holding (> 0-0.5 ha) | | 47% | 22% | |
| Marginal holding (> 0.5-1 ha) | | 20% | 23% | |
| Small holding (> 1-2 ha) | | 21% | 33% | 79%* |
| Semi-medium holding (> 2-4 ha) | | 10% | 19% | 16% |
| Medium holding (> 4-10 ha) | | 2% | 3% | 5% |
| Average size of rubber holdings | 1.13 ha | 0.91 ha | 1.35 ha | |
| Total rubber trees per holdings | | | | |
| 1-100 | | 50% | 28% | |
| 101-200 | | 18% | 20% | |
| 201-400 | | 22% | 26% | |
| > 400 | | 10% | 26% | |
| Average trees per holding | 256 | 190 | 321 | |

*Source:* Author's survey, K.Tharian George (1999: 192).
*Note:* * All holdings up to 2 hectares.

have 0.5 ha or less of total land and 78 per cent own 2 ha or less. These are by definition marginal and small holdings. Semi-medium and medium holdings constitute 12 per cent of the total in Choovoor and 22 per cent in Maravikkallu. Compared to the results of a survey by the Rubber Research Institute of India (RRII) of 2,575 rubber holdings (K. Tharian George 1999: 192), farms in Maravikkallu are of roughly the same size as the average rubber holding in India (79 per cent of holdings with 2 ha or less), while farms in Choovoor are smaller in size. The average total size of the rubber holdings is only 0.91 ha in Choovoor and 1.35 ha in Maravikkallu. Taking both wards together, the average size of holdings is 1.13 ha.

Rubber holdings can possess either immature or mature natural rubber trees or a mixture of both (section 4.2.1). For the purposes of this study, the total size of the rubber plantation is defined as the total number of both immature and mature rubber trees on the holding land at the time of the survey. It is important to note that this definition does not correspond to the normal definition (e.g., of the Rubber Board) of the size of a rubber plantation, in which the size is defined by land area (in hectares). The choice to work according to the number of trees was made on the basis of pre-test interviews, and for three main reasons: (1) the total number of rubber trees takes into account the differences in tree densities that often occur in holdings located in marginal and hilly areas; (2) it can reflect the frequent loss of rubber trees due to storms or diseases; and (3) it is normal practice for many rubber holders to define their rubber plantations in terms of total trees and not by area.

The results of the survey of rubber tree assets on the holdings are shown in Table 5.8. In Choovoor, 50 per cent of all rubber holdings have a total of 100 trees or less, 18 per cent have between 101 and 200 trees and only 10 per cent have more than 400 trees. In Maravikkallu, the holdings are generally bigger: only 28 per cent have 100 trees or less, 20 per cent have between 101 and 200 trees, while 26 per cent of all holdings have more than 400 trees. The average number of rubber trees per holding over the two wards is 256, 190 on average per holding in Choovoor and 321 in Maravikkallu.

Figure 5.4 shows the total number of rubber trees according to the total land size of the holding (in ha) as a graph.

Figure 5.4 demonstrates that there is a correlation between the two variables 'total rubber trees' and 'total land' (Pearson's correlation coefficient = 0.885), as would normally be expected. However, of more interest for this study is the great diversity in the number of trees planted in holdings which have the same total area of land. If one takes the example of all holdings with 1.6 ha of total land (11 cases in total), their number of trees ranges from a minimum of 100 rubber trees (in one case) to a maximum of 500 rubber trees (also in one case), with all other holdings distributed between these two extremes. This is a sign that the income generation of rubber holders in the agricultural sector is quite diverse, some preferring to plant rubber while others focus more on other crops (see chapter 6).

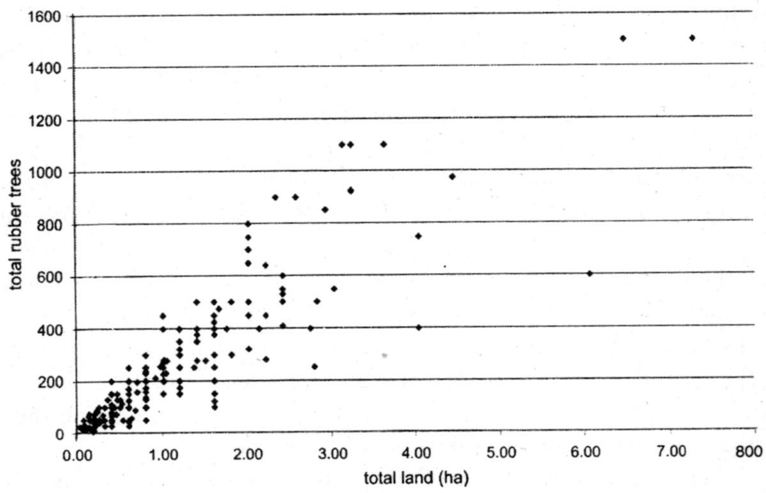


*Source:* Author's survey.

Fig. 5.4: Correlation between total number of rubber trees and total land (in ha) ($n$ = 216)

## *Importance of other cash crops in rubber holdings*

Table 5.9 provides a summary of respondents' answers to the question: Which are the two most important cash crops on the holding besides rubber?

TABLE 5.9: MOST IMPORTANT CASH CROPS OF RUBBER HOLDINGS, BY WARD ($n$ = 216)

| *Cash crop* | *Total* | | *Choovoor* | | *Maravikkallu* | |
|---|---|---|---|---|---|---|
| | *Total* | *%* | *Total* | *%* | *Total* | *%* |
| Coffee | 54 | 25.0 | 36 | 34.0 | 18 | 16.4 |
| Pepper | 44 | 20.4 | 28 | 26.4 | 16 | 14.5 |
| Coconut | 23 | 10.6 | 12 | 11.3 | 11 | 10.0 |
| Cloves (*Gramboo*) | 10 | 4.6 | 10 | 9.4 | – | – |
| Tapioca | 7 | 3.2 | – | – | 7 | 6.4 |
| Vanilla | 5 | 2.3 | – | – | 5 | 4.5 |
| no cash crops/ no answer | 71 | 32.9 | 18 | 17.0 | 53 | 48.2 |
| Total | 214 | 100.0 | 104 | 100.0 | 110 | 100.0 |

*Source:* Author's survey.

Overall, coffee, pepper and coconut are mentioned as the most important cash crops besides rubber.[14] However, results show a big difference between the two wards. Choovoor has a higher percentage of rubber holdings with coffee and pepper as important cash crops than Maravikkallu. This could be a sign of higher on-farm diversification in Choovoor ward. Clove (*Gramboo*) is another crop the survey identified as being important, but mainly in Choovoor. There are, however, other cash crops mentioned as being of economic importance, especially in Maravikkallu, namely, tapioca and vanilla.

The answer 'no cash crops' in Table 5.9 is also of interest. Although this answer can unfortunately not be separated from 'no answer', feedback from the surveyors shows that many rubber holdings have no other important cash crops besides rubber. This is particularly true of Maravikkallu, where a high percentage of holdings (40-5 per cent) state that they have no other cash crops. In Choovoor, the percentage is much lower (10-15 per cent). This suggests that rubber holdings in Choovoor generally focus more on mixed farming practices than those in Maravikkallu; or, put differently, that Maravikkallu rubber holdings are more specialized in natural rubber.

## NOTES

1. For political-administrative purposes, Thalanadu *Grama Panchayat* is part of the *Block Panchayat* of Eratuppetta, which is part of the *District* of Kottayam. For revenue collection affairs, Thalanadu lies in the *Revenue Village* of *Poonjar Vadakkekara* (Poojnar North), which is part of the *Taluk* of *Meenachil.*
2. Approximate figures based on data from Thalanadu Grama Panchayat Development Report (Thalanadu Grama Panchayat 2002, translated into English).
3. Due to an administrative regulation, wards of a Panchayat cannot comprise more than 1,000 voters. If the number exceed the ceiling, then wards have to be restructured. In Thalanadu Panchayat the revision took place between September and December 2003. The ten wards now are Choovoor (1), Chonnamala (2), Meladukkam (3), Adukkam (4), Vellany (5), Chamappara (6), Teekoy estate (7), Maravikkallu (8), Thalanadu centre (9), with Panchayat office) and Ayyampara (10).
4. This information is based on statements by farmers regarding the soil characteristics on their farms.
5. According to interviewees, at that time the land belonged to the Raja of Poonjar.

The first settlers were from wealthy families and bought 100-200 acres, at a price of around 50 cents per acre.

6. An elderly interviewee explained that the fact that British planters managed a rubber plantation with so much care—and were apparently also earning money with it—was reason enough for local people to believe that the crop would have an interesting future. The first farmers who planted rubber collected the seeds in the plantations of the Teekoy Rubber Estates.
7. An informant said that at that time the tapping of 100 rubber trees produced around 2 kg of rubber sheets.
8. The construction of the road was started in 2003, and financed under the Prime Minister's Village Road Development Fund, a food-for-work project.
9. Though government schools are present in the Panchayat, many families send their children to private English schools outside the Panchayat.
10. Agricultural Census of the Indian Ministry of Agriculture (Directorate of Economics and Statistics, various issues).
11. Income from remittances or other transfer payments (e.g., pension) were not surveyed, though it is acknowledged that they can play an important role in certain households.
12. The word *plantation* can be misunderstood, as it is used in different ways. *Plantation* can refer to holdings having more than 20 ha, the so-called *estates*. However, the term *plantation* also refers to any plot cultivated with natural rubber, independent of its size. If not specifically mentioned otherwise, plantation will be used in this second sense.
13. For analytical clarity it is important to note that the definition of the size of the holding as surveyed in this study differs from the Rubber Board of India's official definition. The Rubber Board defines the size of a rubber holding as 'a rubber area contiguous or non-contiguous . . . under a single ownership' (Rubber Board 2003a: ix); this means that it does not include any land area belonging to members of a household other than the one that contains natural rubber. The consequences of the Rubber Board's strictly sectoral approach will be discussed in section 8.
14. Coffee and pepper are generally sold to local buyers. Coconut is first retained for home consumption (coconut is used in most local dishes in Kerala), and only the surplus is sold on the local market.

CHAPTER 6

# Income Generation Strategies and Diversification in Rubber Holdings

This analytical chapter elucidates the income-generation strategies and the role of diversification in rubber holdings of the two wards. Section 6.1 describes the major rubber cultivation practices in Thalanadu, as well as the advantages and disadvantages of the different practices mentioned by rubber holders interviewed in Thalanadu. Section 6.2 details the income analysis of the holdings and suggests a typology of rubber holdings. Based on the typology, the next sections describe the income strategies of marginal holdings with low incomes (section 6.3), of small holdings with medium income (section 6.4), and of small to medium holdings with good to very good income (section 6.5). Section 6.6 describes the impact of and reaction to the 'rubber crisis' between 1997-8 and 2001-2. Section 6.7 analyses in detail the role of income diversity and diversification for each type of holdings, while section 6.8 summarizes the main findings.

## 6.1 RUBBER CULTIVATION PRACTICES IN THALANADU

This section describes the local cultivation and management practices of natural rubber. The main advantages and constraints of cultivating rubber are also listed. The aim of this section is to be able to situate the other income-generating activities in relation to the characteristics and specificities of natural rubber.[1]

### 6.1.1 RUBBER CULTIVATION AND MANAGEMENT PRACTICES

The life cycle of natural rubber can be divided into four different phases: the immature phase during (re)planting, the mature phase, the slaughter-tapping phase prior to logging and the cutting phase when the trees are cut and sold, and land is prepared for replanting.

*(Re)planting and immature phase*

In Thalanadu, replanting is done in September towards the end of the south-west monsoon season, since rain is needed after the seedlings have been planted. Prior to the establishment of a rubber plantation, the soil needs to be levelled (terraced), especially in hilly areas. Some families who did not terrace their land when they planted rubber for the first time are now facing erosion problems. The preparation of the soil (layering) and the replanting work are done by specialized labourers hired for the purpose. The seedlings planted are generally produced and bought locally; it is said that the region around Teekoy is reputed for the high quality of its seedlings. Some of the farmers interviewed had rubber nurseries themselves as a further income-generating activity on their farms. Most farmers nowadays plant the RRII 105 variety, which is known to generate high yields of latex. Most of the admittedly few farmers planting other varieties complained because they were given bad advice and therefore did not plant RRII 105. The spacing of trees depends on the variety planted and is said to be a significant yield variable. According to farmers, older local varieties required greater spacing, while, with the new varieties, spacing can be reduced. However, as some farmers plant mixed-subsistence crops in between rubber trees, they sometimes do not follow the Rubber Board's spacing recommendations (section 6.3).

Rubber seedlings are said to be very sensitive, especially in the first two years after plantating. One problem faced by farmers is the high rate of seedlings lost during this phase, especially to drought or wind. The local survey revealed that the loss of trees in the first seven years is between 15 and 30 per cent. To limit this damage, some marginal farmers, who planted rubber in their mixed plots, replanted each seedling as soon as it died, with the result that they had many trees of varying ages. However, each tree proved important for their income.

Tapping starts, on average, after a period of seven years, as soon as the girth of the rubber trees has reached a certain size. However, some holders start tapping after only five years. They said that they could not afford to wait for two more years before getting some income from the trees. However, many recognized that this was not beneficial for the life of the trees. The lack of monetary income from their own land is a problem for many holders during this seven-year phase. Some do get a certain amount of income by intercropping cash crops amongst the rubber trees: these crops receive much sunlight during the initial years. Some of the farmers interviewed said that they had intercropped mainly banana between the rubber seedlings. In other cases, farmers leased out their land to bigger farmers within the locality, or to business people who were active as pineapple planters.[2] This allowed them to get some income even during replanting.

### *Mature phase: Tapping and rubber sheet processing*

The mature phase starts as soon as the rubber trees are tapped for the first time and lasts until they are cut. After the first tapping, trees are tapped regularly. As Table 6.1 describes, tapping is carried out exclusively by household members in around two-thirds of the cases, while in bigger holdings external tappers are employed. Tapping is a specialized skill. Some rubber holders mentioned that only around 10 per cent of the tappers in the area have the level of skill that is required for excellent tapping and it was often said that it is difficult to employ these tappers nowadays. The employment of a skilled tapper is essential because in plantations in which tapping is done unprofessionally, yield is said to be much lower than the full potential. Although the Rubber Board offers tapping classes in the Panchayat (section 7.1.1), most holders said that they had learnt the skills on the job.

TABLE 6.1: ORIGIN OF TAPPERS ($n$ = 40)

| *Type of tapper* | *From household* | *External tapper* | *Both* | *Not tapping* | *No answer* |
|---|---|---|---|---|---|
| Total | 27 | 8 | 2 | 1 | 2 |
| Per cent | 68% | 20% | 5% | 3% | 5% |

*Source:* Author's interviews.

TABLE 6.2: TAPPING SYSTEM ($n$ = 40)

| *Tapping system* | *Daily* | *Alternate days* | *Each third day* | *Variable* | *No answer* |
|---|---|---|---|---|---|
| Total | 1 | 25 | 3 | 6 | 5 |
| Per cent | 3% | 63% | 8% | 15% | 13% |

*Source:* Author's interviews.

A tapper taps on average around 250 to 350 trees per day. The daily minimum is approximately 100 trees, the maximum around 400 trees. In 2003-4, wages varied between Rs. 25 and 35 per 100 trees and tapping a day. Thus, a person tapping 300 trees in a day will get around Rs. 100 as daily salary. Normally, this will include the processing of latex into the rubber sheets. If rubber plots are far away from the farm and latex or rubber sheets have to be carried to the farmhouse, the salary is higher. When the weather conditions are too extreme (too hot, too windy, too rainy), tapping may not happen. On such days, employed tappers do not get any salary.

Table 6.2 shows that the most common tapping system in the two wards under study is the alternate one, i.e. tapping every second day (d/2). However, some holders mentioned that tapping every third day is more economical. In fact, the total yearly yield of the rubber tree in the long term is equal using either system but, with the latter system (every third day), labour costs can be reduced. Also, tapping every third day is said to reduce the vulnerability to the 'tapping panel dryness' disease.[3] However, some farmers employing tappers said that their tapper also tapping other plots alternately means they cannot change the tapping system. Table 6.3 shows the activity profile of a tapper tapping around 300 trees. As will be seen in other sections, the job of a tapper is quite strenuous and stressful, compared to the job of a 'normal' agricultural labourer.

TABLE 6.3: ACTIVITY PROFILE OF A TAPPER TAPPING AROUND 300 TREES

| *Time* | *Activity* |
|---|---|
| 4.00 a.m. | Wake up |
| 4.30-9.00 | Tapping (cutting) 300-350 trees (from 6.30-8.00 if only 100 trees) |
| 9.00-9.30 | Eat breakfast (in the field, in own house or in employer's house) |
| 9.30-12.00 | Collection of latex from trees |
| 12.00-2 p.m. | Production of rubber sheets |
| 2.00-2.30 | Lunch (in own house) |
| 2.30-5.00 | Kooli (labourer) work, if available (0-3 times per week) |
| 5.00-8.00 | Free time (gather in common place and talk about village issues) |
| 8.00 | Eat dinner, then go to bed |

*Source:* Author's interviews and observations.

The tapping season unfolds as follows: the peak season with the best yields is in October and November. In December and January, yield is dependent on rainfall. Most planters stop tapping by the end of January and until the beginning of April. This is called the 'wintering season', when trees lose their leaves. The months of April, May and part of June provide good yields, but this depends on the local climate. Many stop tapping during the south-west monsoon rains in June, and start again in August or September. With alternate tapping, each tree is tapped on average of 80 to 120 times per year. Rainguards[4] are used to extend the yielding season. Holders using rainguards said that they managed to have 120 to 160 tapping days per year. However, if the rains are too heavy, even rainguards cannot prevent tapping from being interrupted. Farmers tend to care less for their rubber trees at a later stage. The important thing at this stage is to collect the maximum amount of latex the trees are still able to provide and thus the tapping rhythm is often increased to a maximum.

The farmers interviewed repeatedly mentioned two important constraints of plantations in Thalanadu. Firstly, the wards studied seem to be frequently affected by storms, especially the hillier ward (Choovoor) which is less protected. Along with diseases and pests, which also affect individual trees, the annual loss of trees can be considerable on certain holdings. Secondly, many farmers mentioned that they did not use the amount of fertilizer prescribed by the Rubber Board, and thus expected lower yields. However, the use of fertilizer is very dependent on the rubber price. In some cases, local cow dung is used as a substitute for chemical fertilizers.

In most smallholdings, latex is transformed into RSS rubber sheets. Latex is processed with either the holder's own sheeting roller, or his/her neighbour's.

### *Slaughter-tapping and cutting phase*

The last phase in the tapping of a plantation, called the slaughter-tapping phase, is carried out either by farmers themselves or by other people who lease the land and rubber trees of farmers for a period of two to three years.[5] During this time, they will use special tapping techniques to extract as much latex as possible from the trees. 'Slaughter' refers to the fact that the trees are slowly 'killed' by these techniques. When this type of tapping is over, trees are felled by professional loggers. The income a farmer gets from rubber wood depends on the demand and the use of the wood (i.e. as firewood or for furniture). In fact, when plantations are located in remote places without access by road, it can be difficult to transport the logs and the tree is then sold in small pieces as firewood. In very remote places, only the branches are cut and burnt, and the trunk is left as it is. This gets rid of the shade and the land can be replanted with other crops. In other cases, farmers use the wood to produce charcoal on the land before selling it. However, this is not economically viable.

### 6.1.2 Advantages and Disadvantages of Rubber Cultivation

Based on interviews with rubber holders, characteristics which some farmers judge to be an advantage are mentioned by others as a disadvantage. This study reproduces these comments as they light on the different perspectives of different rubber holders.

The biggest advantage mentioned by almost all holders is the regular yield (and thus income) from a rubber plantation. As mentioned above, rubber can yield between 80 and 120 times per year, and even more times when rainguards are used. This is a big advantage of rubber compared to other crops. Regularity of income is considered important, even when the rubber price is comparatively low. Thus, the fact that a regular income is available seems to be a more important aspect, especially for marginal holders, than the total income over the year. Furthermore, rubber provides a long-term income with a good yield for over twenty years, and its price is no less stable than that of other crops. A further advantage of rubber is the fact that the cost of a rubber replanting can be covered with the income from the sale of wood from the previous rubber plantation.

Another important advantage is that rubber does not require a high annual workload once tapping has started. Manuring, clearing land and some irregular spraying against pests are the main tasks, besides tapping. Although the costs of tapping can be substantial, the fact that one's own household can do it reduces monetary expenditure to a great extent. Thus, the substitution of employed labour by family labour is an advantage, especially when the rubber price is low. Rubber does, however, require much attention during the replanting phases, especially in the first two years. This was cited as a disadvantage, especially by the marginal holders.

Other advantages of rubber are its general resistance to the climate and to diseases and that it can be planted even on rocky soils. These advantages have, however, all to be considered in comparison to other crops planted in the region. For example, the seasonal drought that affected the region during the field study proved that, while a crop like pepper was rapidly affected by the lack of water, rubber still coped quite well; though it will yield less, only an extremely severe drought would affect the rubber tree in the long term. Rubber also does not require irrigation (i.e. from rainwater) in the same way as, other crops.

The main disadvantage, mentioned is that rubber gives its first yield only several years (six to seven) after it has been planted. For some holdings, this income gap represents a problem since other income opportunities have to be sought during this time. A plantation is also quite cost-intensive. Another problem is that of the seasonality of latex yields. During the months of February, March, and from July to September, tapping is reduced or even discontinued. Also during this time, many holdings have to seek alternative income sources. And there is the difficulty, for bigger holders, of finding skilled tappers.

The fact that rubber plantations generate a lot of shade was mentioned as a constraint by many holders. On the one hand, it is difficult to plant other crops *within* a rubber plantation, and, on the other, land in *between* rubber plantations is often too shady, thus reducing the potential yield of plants. Some interviewees also mentioned the lack of light and the lack of a 'clear atmosphere' as sometimes having an impact on the morale of household members. Some mentioned the lack of food trees (e.g., coconut) and crops around the homesteads due to the presence of rubber plantations.

Finally, some comments were made about the negative long-term impact of fertilizer on the soil quality in rubber plantations; others complained about the cost of fungicides which need to be applied regularly so as to protect the trees from specific diseases. It was mentioned that buying and applying fertilizers and fungicides is only possible when the price of rubber is not too low. As already mentioned, there is the further constraint of storms regularly destroying part of plantations, especially in the hilly areas.

## 6.2 INCOME ANALYSIS AND TYPOLOGY OF RUBBER HOLDINGS

Based on the results of the household survey in the two wards, in-depth interviews were done with forty rubber holdings chosen following a random sampling. These interviews included a detailed survey on the gross income of the households during the year 2003-4. Section 6.2.1 describes and analyses the results of the survey on the gross annual income. In order to provide a more detailed analysis of the income strategies, a typology of rubber holdings is developed and the different types of holdings described in section 6.2.2. The different types of holdings form the basis for the detailed analysis of the income strategies in sections 6.3-6.7, and their links to the local institutional setting (Chapter 7).

### 6.2.1 Income Analysis of Rubber Holdings

The yearly gross income of the household including all income activities of all household members during the financial year 2003-4 was taken into account while doing an income analysis.[6] The annual gross total income is composed of the sum of all incomes of a household without deducting either the costs of these activities or the taxes. Thus, it does not correspond to the profit of the household.

Income data was categorized following the classification suggested by Christopher Barrett et al. (2001).[7] For analytical purposes, households were first separated according to the total holding size, following the official size categorization and denomination of the Agricultural Census of the Indian Ministry of Agriculture (Directorate of Economics and Statistics, various issues): *marginal* holdings (<1 ha), *small* holdings (1-2 ha), *semi-medium* holdings (2-4 ha) and *medium* holdings (4-10 ha).[8] Then, in a further step, households of each size category were separated into two groups according to the importance of agricultural and non-agricultural incomes. The first group of holdings base their income on agriculture, i.e. the households have more than 50 per cent of total income derived from the agricultural sector. The second group of holdings base their income on non-agricultural activities, i.e. their households derive less than 50 per cent of total income from the agricultural sector. Table 6.4 shows the results of this classification.

A look at the total income of the different holding categories (Table 6.4) shows that there is great heterogeneity between the incomes of rubber holdings of different sizes. In fact, while the average yearly gross income of *marginal* holdings (< 1 ha) is Rs. 29,215, the *small* holdings category (1-2 ha) shows an average income of Rs. 131,904 although the holdings are, on average, a maximum of twice the size of the marginal ones. In the *semi-medium* (2-4 ha) category, the average yearly gross income is Rs. 393,526, three times the previous category. This suggests an exponential increase in the total income when the size of the holding grows. Furthermore, an analysis of the difference between agricultural and non-agricultural incomes is of great interest for this study. Table 6.4 shows that, for all size categories, agriculture-based holdings always derive more than 80 per cent of their total income from agriculture. In two cases (*semi-medium* and *medium* holdings), this share even reaches around 90 per cent. This suggests that agriculture-based holdings are highly dependent on the agricultural sector for their income. On the

TABLE 6.4: AVERAGE GROSS INCOMES (IN RS. PER YEAR) OF HOLDINGS ACCORDING TO SIZE AND MAIN SOURCE OF INCOME (AGRICULTURE AND NON-AGRICULTURE) ($n$ = 40)

| *Size category of holdings* | *Total average gross income (in Rs.)* | *Part of agricultural income (in Rs. and %)* | *Part of non-agricultural income (in Rs. and %)* | *Total agricultural income/ha (in Rs.)* |
|---|---|---|---|---|
| Marginal (<1 ha) | 29,215 | | | |
| agric. based | 21,824 | 17,443 (80%) | 4,381 (20%) | 34,397 |
| non-agric. based | 48,431 | 11,164 (23%) | 37,267 (77%) | 38,853 |
| Small (1-2 ha) | 131,904 | | | |
| agric. based | 106,153 | 84,600 (80%) | 21,553 (20%) | 60,438 |
| non-agric. based | 189,842 | 69,808 (37%) | 120,034 (63%) | 48,590 |
| Semi-medium (2-4 ha) | 393,526 | | | |
| agric. based | 419,658 | 381,121 (91%) | 38,536 (9%) | 140,037 |
| non-agric. based | 236,734 | 35,787 (15%) | 200,947 (85%) | 16,048 |
| Medium (4-10 ha) | 643,073 | | | |
| agric. based | 643,073 | 563,059 (88%) | 80,015 (12%) | 103,061 |
| non-agric. based | n.a. | n.a. | n.a. | n.a. |

*Source:* Author's interviews.

*Note:* Exchange rate (May 2004): 1 Euro = 56 Rs. (Indian Rupee, INR).

contrary, holdings depending on non-agricultural income still also rely on income from agriculture. This is especially so among *marginal* holdings, whose agricultural income is often based on irregular wage employment. It is interesting that there is no case amongst the *medium* holdings (4-10 ha) of a farm which bases its income on non-agricultural activities. The hypothesis is that, as they derive a fair income from rubber, there is less scope for non-agricultural activities.

The last column of Table 6.4 shows calculations of the average total agriculture income per hectare. As expected, this income is higher in agriculture-based holdings than in non-agriculture based ones for all categories except marginal holdings. This deviation can be explained by the fact that the agricultural income is derived from off-farm wage labour in the agricultural sector; however, this income is not bound to one's own farmland and thus distorts the average agricultural income per hectare. The fact that average income per hectare increases in line with the size of the holding is an interesting element of the results for the agriculture-based holdings. While it amounts in marginal holdings to around Rs. 34,000 per ha, in the smallholdings it is around Rs. 60,000 per ha and, in the semi-medium and medium holdings it is more than Rs. 140,000 and Rs. 100,000 per ha respectively. This is almost certainly linked to the higher average number of rubber trees in bigger holdings and the high income obtained from rubber during 2003-4. It is important to note that during the time of the field study the price of natural rubber was well above the average price of the years 1997-2001 (Fig. 4.7).

### 6.2.2 Income-Based Typology of Rubber Holdings

A typology of natural rubber holdings has been developed to allow a better understanding and analysis. The typology relies first of all on the differentiation between agriculture-based and non-agriculture-based holdings. It is then complemented by a classification of the yearly gross income, developed on the basis of interviews with rubber holders and local experts: *low* income (< Rs. 50,000 per year), *medium* income (Rs. 50,000-150,000), *good* income (Rs. 150,000-350,000), and *very good* income (> Rs. 350,000). Thirdly, the classification is also based on the analysis of the income strategies of each household (e.g., having a job in government or not, leasing land for agricultural activities or not, etc.). These qualitative factors explain why certain holdings within the same size category (e.g., < 1 ha) can be found in different typologies. Tables 6.5 and 6.6 show the typology of rubber holdings.

In the following paragraphs, the different types of holdings are described in a nutshell, as well as the role of natural rubber in the different holdings. In sections 6.3, 6.4 and 6.5, the income strategies of the different types of holdings will be described and analysed in depth. It is important to note that the inclusion of a holding in one category is based on the status of this holding during the interviews in 2004. However, the income situation of a holding can sometimes change suddenly and to a great extent. The categorization of holdings has thus to be seen as a fluctuating one, depending on upward or downward mobility.

### *Type I: marginal holdings, agriculture-based, low income*

Although holdings type generate some income from rubber, in most cases the holders cannot sustain their livelihood with it alone. Besides rubber, they derive income from the sale of other cash crops. An important part of their income is drawn from wage labour as rubber tappers in

TABLE 6.5: TYPOLOGY OF RUBBER HOLDINGS WITH MAIN INCOME SECTOR, SIZE OF INCOME, AVERAGE TOTAL INCOME, AND AVERAGE INCOME FROM EACH SECTOR ($n$ = 40; IN RS.)

| *Type* | *Size category of holding* | *Main income sector* | *Gross annual income* | *Average total income* | *Avg. agric. Income (in Rs.)* | *Avg non-agric. income* |
|---|---|---|---|---|---|---|
| I | Marginal (<1 ha) | agric. | low | 21,824 | 17,443 (80%) | 4,381 (20%) |
| II | Marginal (<1 ha) | non-agric. | low | 48,431 | 11,164 (23%) | 37,267 (77%) |
| III | Small (1-2 ha) | agric. | medium | 106,153 | 84,600 (80%) | 21,553 (20%) |
| IV | Small (1-2 ha) and Semi-medium (2-4 ha) | non-agric. | good | 199,221 | 63,004 (32%) | 136,217 (68%) |
| V | Semi-medium (2-4 ha) and Medium (4-10 ha) | agric. | very good | 475,512 | 426,606 (90%) | 48,906 (10%) |

*Source:* Author's interviews.

TABLE 6.6: TYPOLOGY OF RUBBER HOLDINGS WITH AVERAGE LAND SIZE, TOTAL RUBBER TREES, RUBBER TREES PER HA, INCOME PER MATURE TREE, AND MAIN INCOME ACTIVITIES ($n$ = 40)

| | *Avg. total land (ha)* | *Avg. total rubber trees* | *Avg. rubber trees per ha* | *Avg. rubber income per tree* and year* | *Main income activities (in order of importance)* |
|---|---|---|---|---|---|
| I | 0.51 | 88 | 173 | 133 | Agricultural wage employment (tapping, daily labour work)<br>Rubber & mixed cash crops<br>Non-agricultural self- and/or part-time wage employment |
| II | 0.29 | 51 | 176 | 179 | Non-agricultural self- or wage self-employment (in government or private sector) Rubber |
| III | 1.40 | 316 | 226 | 246 | Rubber income<br>Cash crops (if available) |
| IV | 1.59 | 366 | 230 | 173 | Full-time wage employment (in government or private sector) or attractive self-employment<br>Rubber |
| V | 3.41 | 971 | 285 | 297 | Rubber (specialized rubber planters)<br>Cash crops on leased land |

*Source:* Author's interviews.
*Note:* * only mature trees.

larger holdings, or from daily wage labour income from agricultural work on other farms. In some of the holdings, household members have either part-time wage employment in local factories or they pursue part-time self-employment at home (e.g., stitching, or some small trade business). Through the combination of all these activities most of them can make a basic living. Most marginal holdings of Type I are often found in the hillier part of Choovoor, 'upper Choovoor' (Fig. 5.1). The share of households belonging to the Scheduled Tribe and Scheduled Caste communities is higher within Type I than in the other holding types.

### *Type II: marginal-size, non-agriculture-based, low income*

Holdings of Type II–as those of Type I—provide income from rubber and a few other cash crops, and from agricultural wage labour. However, this income is not sufficient for the holders to make a living. They get their main income from off-farm wage employment (e.g., working in a medical shop, in a self-help group shop or—as seasonal migrants—in a T-shirt making factory) or from non-agricultural self-employment on the farm (e.g., preparation of Ayurvedic medicine) or off-farm (as an Ayurvedic doctor). The fact that they derive a better income by investing their time in these non-agricultural opportunities leads to a situation where rubber is often managed in a sub-optimal way.

*Type III: small-size, agriculture-based, medium income*

The number of rubber trees in the Type III holdings is higher than in the two types described above. As a consequence, they provide a good income from rubber and most households manage to cover their household expenses with that income. This income is sometimes complemented by sales of other cash crops, including some planted on temporarily leased land. Members of these holdings do not work as wage labourers or tappers on other farms; on the contrary, due to their big rubber plantations, they act as employers for tappers from outside their households. Geographically, these holdings are mostly in Maravikkallu ward.

*Type IV: small and semi-medium size, non-agriculture-based, good income*

The total number of rubber trees in Type IV holdings is similar to that in Type III (Table 6.6). This gives the holders a certain income, complemented by income from other cash crops. However, their main income comes from the non-agricultural sector. In fact, they are either full-time employees—often in civil service jobs—or self-employed as doctors, taxi drivers, tea shop owners, or other. Though income from rubber is important to cover basic household expenses, they have to devote their time to their non-agricultural jobs and employ labourers to look after the rubber plantation.

*Type V: semi-medium and medium size, agriculture-based, very good income*

Rubber holders with extended plantations specialized in rubber cultivation fall in the Type V category. As a secondary income, most of them were found to be involved in important seasonal cash crop cultivation, mainly banana and pineapple. Interestingly, these activities take place on leased land. Only in a few of these holdings do household members engage in the non-agricultural sector, and 90 per cent of their main income is drawn from agriculture. All of these holdings, except one, are located in Maravikkallu ward.

Figure 6.1 summarizes in graph form the different typologies of rubber holders according to the size of their rubber plantation (total number of trees) and their gross annual income.

### 6.2.3 Comparative Analysis of the Role of Rubber

As expected, the average number of rubber trees per hectare of total land is lowest in holdings of Types I and II. This can be explained by the fact that these marginal holdings are in hilly areas where rubber trees are planted less densely, but also by the importance of a farming system based on mixed cultivation (section 6.3). The situation is different in holdings of Types III and IV which have a higher average number of

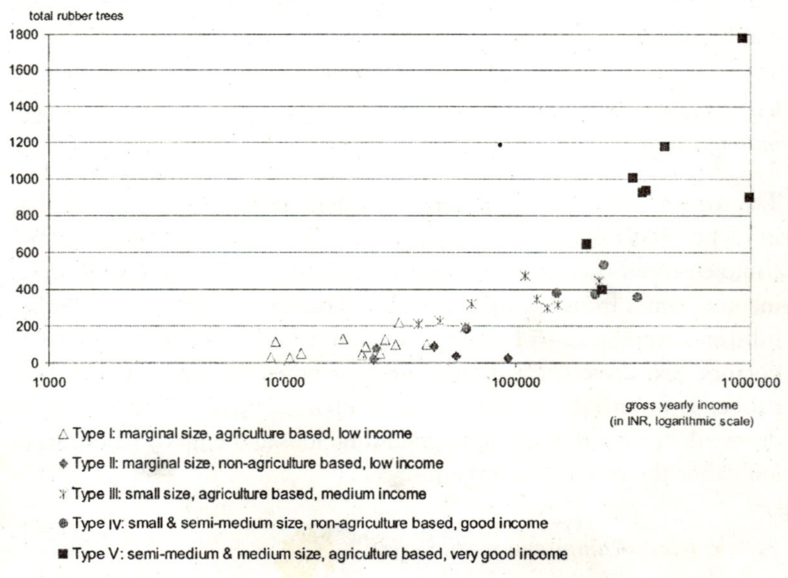


*Source:* Author's interviews

Fig. 6.1: Typology of rubber holders in Choovoor and Maravikkallu wards ($n$ = 40)

trees per total land. The agriculture-based holdings (Type III) do in fact cultivate rubber as their main activity, while the non-agriculture-based holdings (Type IV) mostly plant their land with rubber which is fully managed by external labourers, sometimes at a suboptimal level due to lack of supervision. Thus, concerning rubber, the main difference between the two types lies more in the intensity of rubber management practices than in the actual size of the plantations.

Data on the average annual income per mature rubber tree (Table 6.6) backs up the finding that holdings of Types III and V are the ones deriving the greatest income from rubber (with Rs. 246 and 297 per tree and per year, respectively). Again, these are the holdings that depend to a great extent on income from rubber. The low income per tree and year for the holdings of Type I (Rs. 133) can be explained by the often unsuitable land on which rubber is planted (i.e. steep and rocky) and, to a certain extent, by plantation management practices that are less developed than the ones in bigger holdings. What is also clear from this data is that holdings of Types II and IV, which receive most of their income from non-agricultural activities, do not get as much income from rubber as the other types. In holdings of Type IV, this is also due to the lack of supervision.

## 6.3 INCOME STRATEGIES OF MARGINAL HOLDINGS WITH LOW INCOME

This section details the income strategy of the marginal holdings (< 1 ha) with a low total income (< Rs. 50,000 per year). The first part is dedicated to holdings of Type I which derive most of their income from the agricultural sector. The second part details the income strategy of holdings of Type II which derive most of their income from non-agricultural activities.

### 6.3.1 Type I: Income from Own Rubber, Other Mixed Crops, and Agricultural Wage-Employment

The combination of rubber, other mixed (cash) crops and agricultural wage-employment is the main income 'strategy' of holdings of Type I. As seen in Table 6.6. The average of 88 rubber trees and an average total land holding of 0.51 ha is insufficient to cover the household expenses. Thus, the households engage in different activities, some on-farm, others off-farm. It is important to recall that only with the combination of these different activities can the necessary means to ensure the livelihood of these holdings be provided.

*Diversified agriculture on own farm*

Though rubber does not guarantee the necessary income for these farms, it still accounts for an important share of total income. In addition, the households often mention the advantage of rubber providing a steady income (section 6.1.2). Their own rubber is all tapped by household members who have all learnt tapping on-the-job. Since they often have wage-employment as tappers on another, bigger rubber farm, the tapping of their own trees is dependent on the time that remains after their hours of employment. This sometimes leads to a situation where their rubber is only tapped when there is a free day, i.e. irregularly.[9] There are, however, other reasons why personal rubber plots are managed suboptimally. Since rubber trees are often grown within mixed farming plots, there is sometimes too much shade, which limits the growth and production of the tree. This has also caused some farmers to plant rubber with different spacings than those suggested by the Rubber Board field officer. Furthermore, the pressing need for cash of some farmers leads to a situation where rubber trees are tapped prematurely, after the fifth year of growth; they are also tapped during the wintering season, when tapping activities normally cease (section 6.1.1). Finally, the lack of cash prevents most holders from applying fertilizer, except for some irregular cow dung from their own cow(s), if they possess any.

As mentioned above, marginal holdings follow a mixed cropping strategy. Perennial cash crops such as coffee, cocoa, arecanut, coconut, clove, nutmeg, and sometimes trees for lumber such as jackfruit, teak or anjili are grown beside, and sometimes within, rubber tree plots. In some cases, seasonal staple foods such as banana and tapioca are grown on locations what are unsuitable for rubber (e.g., around the house or on rocky spots). Furthermore, some farmers cultivate vegetables around the house, and collect leaves from wild plants. Holders mention the following reasons that drive them to plant mixed crops (not in any order of importance):

- Food production, to be less dependent on markets
- Regular income, since some crops can be sold at most times of the year (different seasonality of crops)

- Less dependency on the price of only one or a few crops
- Belief that planting only rubber has a negative effect on soil fertility
- Personal motives such as the attachment to multiple crops grown on own farm
- Unclear property relations which make it impossible to take the risk of investing in planting of rubber trees[10]

An important reason is also the remote geographical location of most of these holdings. Since daily food markets are distant, the production of subsistence food crops is cheaper than buying fruits and vegetables on the market.

The planting of mixed crops, however, also has some constraints. Firstly, managing a mixed-crop garden professionally requires the full involvement of one or more people for the day-to-day activities (watering, gathering, marketing, etc.). In seasons when these people can access agricultural off-farm wage-employment at a higher income, their own mixed farms are often managed suboptimally. Therefore, such opportunity costs have to be regularly re-evaluated. Secondly, the shade created by the many tall rubber trees limits the growth of other mixed-food crops (and vice versa). Unless farmers enhance the spacing between the rubber trees or create spots where they choose not to plant rubber—the latter often being the case around houses—mixed crops do not give the expected yield. Finally, farmers who plant mixed crops within their rubber plantation do not have the right to get (re)planting subsidies from the Rubber Board, as the Rubber Board does not permit such practices (section 7.2.1). While this is sometimes a constraint for some farmers, others 'strategically' make this choice, since they expect more income from trees for lumber and other crops (multi-cropping) than from the Rubber Board subsidies. The workings and impact of the Rubber Board subsidies are detailed in section (7.1.2).

Income from livestock and animal rearing, mostly milk production from goats and cows and meat production from pigs, does play an important role in some farms. Livestock rearing was mentioned as being important for the regular production and application of cow dung on farmers' own land as well.

*Agricultural employment as tappers and agricultural labourers*

For half of the interviewed farmers of Type I, income from agricultural wage-employment represents the second pillar of their household income

(Table 6.7). Most of them are employed as tappers on a regular basis, but also have irregular employment as daily wage labourers (known locally as *kooli*).

TABLE 6.7: AVERAGE GROSS INCOMES (IN RS. PER YEAR) OF TYPE I, PER SECTOR ($n$ = 13)

| *Size category* | *in Rs. per year* | *in %* |
|---|---|---|
| Agriculture, on own farm | | |
| from rubber | 8,453 | 39% |
| from other crops | 3,314 | 15% |
| Agriculture, wage employment | | |
| from rubber tapping | 3,188 | 15% |
| from agricultural labour (*kooli*) | 2,488 | 11% |
| Non-agriculture | | |
| from external employment (e.g., in factory) | 1,806 | 8% |
| from self-employment at-home (e.g., stitching, rubber sheet production) | 2,575 | 12% |
| Total | 21,824 | 100% |

*Source:* Author's interviews.

Employment as a tapper is normally fixed for one year and, if no problems arise, it is continued the following season. However, the tapper only earns on the days he taps. Since the tapping season only lasts between eight and nine months a year, this income is missing in the winter season. Also, income during the rainy season depends on the owner's ability and willingness to invest in rainguards. If not, earnings during the rainy season are heavily reduced. This is an important consideration for tappers and their households. In Thalanadu, a tapper's salary depends on his/her bargaining power with the employers; normally it is Rs. 25-35 per 100 tapped trees.[11] The number of trees tapped averages 250-300 per tapping day; these trees are normally tapped alternately (every second day). If the plantation owner has 400-700 trees, it can happen that the same tapper works every day on a particular farm, tapping half of the total plantation (200 to 350 trees) each day. But in plantations with less than 300 trees, a tapper will work only every second day. A tapper's decision whether to work every day also depends on the size of his own rubber plantation. Box 6.1 gives a concrete example of the work of a tapper combining tapping in his own plot and being employed as tapper.

### Box 6.1: Combining Work on One's Own Farm with Employment as a Tapper

Mathew[12] lives in the hillier part of Thalanadu together with his wife, their 3 children and his parents. Some 20 years back, the family bought a small rubber plot on which there are now around 100 trees. Of these, 70 have to be tapped regularly while 30 are still too small. Since the family does not manage to make a living from the latex they get from their plantation, Mathew works in two plantations in the neighbourhood belonging to two different owners. One day he taps 400 trees in a plantation near his house, the following day he taps first his 70 trees, then 200 trees belonging to the second owner. After having tapped the trees, he has a quick breakfast, and then he has to collect the latex and take it uphill to the house, where later in the afternoon, he will make the sheets. Since the trees produce a lot of latex and the plots are very steep, the work is very hard. Still, it has to be done and is part of a tapper's job. However, his work becomes strenuous when, because of a rainy day, he has to postpone the tapping of a plot to the following day. Then he has to tap 670 trees in one day. Usually, in that case, he skips his own trees, but to tap and collect latex from 600 trees is a huge job. He literally has to run to finish in time!

After 20 years of tapping, he is bored with his work, besides having some problems with his shoulders. He would like to change his job, especially because tapping, collection and sheeting together is hard work. Some years back, merchants would collect the latex directly and the tapper did not have to make the sheets. Another problem with tapping is that a tapper earns nothing during the rainy season when plantation owners stop tapping.

During the interviews with households of which a member worked as an employed tapper, it was often stated that it was difficult to find a job, mainly because many farmers did the tapping themselves (Table 6.1). Though on average tapping one's own plot is very frequent in the wards compared to other regions in Kerala, larger farmers seeking to employ tappers mentioned the difficulty of finding skilled and trustworthy employees. The truth must lie somewhere in-between. This was confirmed by the fact that despite there not being many potential employers, tappers with excellent skills and good past records can find employment quite easily. When the collaboration between a tapper and his employer is satisfactory to both, it usually lasts for a long time. This brings advantages for both, as the employer and the tappers enter into an interlocking relationship.

The combination of external tapping with work on one's own rubber plantation (Box 6.1) is possible because tapping activities take place from early morning until early afternoon, when the rubber sheet preparation is finalized (Table 6.3). Thus, in the afternoon, there is spare time that many tappers use for further income activities, mainly *kooli* activities. These activities are either on the same farm on which the farmer is employed as tapper, or on different, neighbouring farms. Nevertheless it is not easy for tappers to get these jobs regularly. One reason for this appears to be linked to the fact that bigger farmers employing tappers and *kooli* labourers differentiate between somebody 'being a tapper' and somebody being a 'daily agricultural worker' (*kooli*). One reason might be their different social status.[13] Thus, there is a tendency to employ tappers only for tapping and daily agricultural workers for all other agricultural work. For this reason, somebody who works as a tapper can have difficulties in accessing agricultural labour work outside his employer's farm. These employment opportunities therefore remain irregular, and are often linked to other happenings, such as an absent *koolis* who has to be replaced quickly. Irregularity of employment is, however, also due to seasonalities, and this is so for employment of tappers (see above) and for *koolis*: for example, some of the jobs *koolis* are specialized in, such as yearly land clearing, are only seasonal.

A further constraint to combining tapping (in the morning) with other work in the afternoon is the heavy workload and stress. Rubber tapping is characterized by regular stressful situations besides being physically demanding.[14] Thus, even if work is easily available in the afternoon, it remains difficult for a tapper to work as a *kooli* every afternoon. Financial pressure does, however, force certain farmers to work beyond what they can physically stand (see below).

### *Rare and part-time: non-agricultural employment*

Some of these households also have wage-employment in the non-agricultural sector, which is, however, of secondary importance income-wise. Besides making the rubber sheets, which, as has been explained above, is a (non-agricultural) service, jobs in the road construction sector,[15] for example, or irregular wage work in a local tea factory have been cited by household members. Sometimes self-employment in the home does exist (stitching activities for neighbours, or small-scale trade in clothes). However, all these jobs are, again, characteristically irregular and part-time, and sometimes seasonal.

*Vulnerability and 'emergency money'*

Analysis of interviews and observations shows a high vulnerability of marginal farmers of Type I. The main reason these households are in a vulnerable situation is the low income they have to survive on; this income is enough for the day-to-day household expenses but not to make any savings. This is also demonstrated by the fact that these households sell their rubber sheets on a daily or weekly basis; they cannot afford to stock the sheets.

Vulnerability is due to factors both external and internal to the household. The impact of climatic variability (rain, wind, and drought due to climate change as is often mentioned)[16] on the yield—and thus also on the income from—crops was said to be a big external constraint. An example often mentioned was that the holdings' locations are responsible for the regular damages caused to rubber and other crops by storms and drought. However, analysis of the different households shows that internal factors such as serious disease, physical disability or even the unexpected death of a household member are important threats faced by many households. And these are double risks: firstly, there is a loss of income because the member of the family can no longer work and, secondly, the cost of care, including visits to doctors/hospital and the purchase of medicine can become prohibitive, especially in the long-term. In many cases, health problems (backache, sight problems, 'mental tension') are directly linked to their jobs as tappers or as *koolis.* In one case, physical problems led to the disruption of a life that had begun impressively (Box 6.2). Thus, the fact that households lack the capacity to bear risks and that there are no formal insurance schemes represents an important constraint. Certain households spoke of what they called 'emergency money'. By this, they meant the availability of small quantities of stocked cash crops (mostly coffee, pepper or some rubber sheets) to be sold only in an emergency, for example to buy medicine. However, this 'emergency money' is limited. In most cases households had to access funds through informal networks (neighbours), but sometimes also through formal credit schemes (see Box 6.2).

Box 6.2: James' Story of a Sudden and Unexpected Change

James was born into a family of landless labourers. His father was already an agricultural labourer, and James, one of his five children, also became one. As such, James did not get any land when he left his village. However, over the last 20 years, James has worked very hard, both as a tapper in different plantations and as a part-time labourer. Since he proved to be an excellent labourer, he had no problems finding work. His plans were to work as much as possible in order to make enough savings to buy a small plot and build a house before getting married. In 1993, he bought a small piece of land with a house and got married. In 1994, he found a better plot on which to cultivate rubber, which he decided to buy for Rs. 30,000. Since he wanted to build a house on that plot, he worked day and night—for example, he started to tap rubber trees at 2 a.m. so as to tap more trees—and successfully accumulated enough money to buy construction material. During his free time, he first built a terrace on his plot and then started to construct his house. He had to bring bricks, sand and 80-kg bags of cement uphill. Though many friends and neighbours told him to take some breaks, he did not because he wanted to finish his family house. However, during the last years, he often felt that something was wrong with his back: he had regular backaches and some problems while walking. Five months later, he was no longer able to walk and, some weeks later, he had to undergo an emergency surgical operation. To pay for the operation, he borrowed Rs. 40,000 from friends and Rs. 20,000 from a bank. However, these debts were in addition to a loan of Rs. 16,000 he had not yet managed to pay back. Unfortunately, the operation did not change his situation much. Though he was in less pain, he had to stop working as a farmer. But since this was his only source of income and his only skill, he now does not know what to do and feels quite desperate. He would like to rent a small petty trade shop, but he does not know how to get a loan for that. At the moment, he does not even know how to repay his debts since the income from his small plot is not enough to cover the household expenses.

Life-cycle events which incur heavy expenses (e.g., dowries at weddings, funerals) are another reason for some households being placed in a vulnerable situation. An important difference between these events and the unexpected ones mentioned above is that the latter are usually more predictable in terms of time and costs. However, they demand long-term strategies and solutions, as the costs involved are often exorbitant. The fact that there is rubber on these holdings is of some help. Often, weddings (including dowries) are paid for from sales of the lumber when a rubber plantation is felled; sometimes, the money comes from the sale of other large lumber trees. The consequence of this, however, is that this money is then not available to either pay for the costs of replanting rubber, or to cover the household costs before the rubber trees start to produce. As an alternative, one farmer used an agricultural loan from a private bank to pay for the dowry of his sister.

### *Gender and intra-household relations*

Analysis of the role of women in the different types of holdings shows that (Type I) women play a different role in income generation on marginal holdings than they do on bigger holdings. In fact, women often have off-farm jobs in the agricultural sector (e.g., as a *kooli* for clearing land), although they are often paid less than their male counterparts.[17] One hypothesis is that women can make use of these opportunities because of the lower status of the family. This would not be possible—or is not practised—by women in households with a high social status (section 7). However, it is also important to mention that, in specific cases, some family members (often the husband) do not allow women (often the wife) to take an off-farm job, although this would have been of great help to the household. There are some advantages to having more household members—normally the wife and husband—engaged in paid work. Firstly, it represents an additional source of cash income for the household. Secondly, it spreads the risk of income failure amongst more than one member and this can be critical in some circumstances. However, the role of women related to on-farm agricultural work is also important: in some cases, women take over the (hard) job of collecting latex from the plantation after their husband has tapped the trees.[18] The advantage is that the husband is then freed from this work and able to start another wage employment earlier in the day. Despite this advantage, some women mentioned that their multiple daily activities were very stressful since they spent the whole day running from one task to the next. For example, one woman was involved in household work, collecting latex and cattle-rearing (including milking and feed collection), besides her business work as a private cloth trader.

*Geographical marginality*

It is interesting to note that many Type I households are situated in rather marginal zones, mostly in the hilly areas of Choovoor. The remoteness, distance and the difficulty of accessing main centres are seen by most farmers as limiting factors. Added to this is the fact that some hilly and rocky areas in Choovoor are facing heavy soil degradation; there have also been recent incidents such as falling rocks causing heavy damage to houses and agricultural plots, thus creating a sense of insecurity. Geographical marginality also implies high transport costs[19] to reach the sites of wage employment[20] or to sell one's products in nearby centres, as well as other costs such as the spoilage of agricultural products during transportation. Geographical distance is also a constraint when people need to settle administrative matters (e.g., joining agricultural schemes) at offices which are sometimes located within the Panchayat, but more often outside it. The costs of transportation and the time spent reaching these offices are high; indeed, if administrative staff create problems (e.g., by asking for bribes), then these costs can become prohibitive. This was the reason why one household abstained from requesting for agricultural subsidies. Another constraint when participating in subsidy schemes, such as the one for the construction of houses, is that the geographical remoteness of the applicants is not taken into account. One interviewee explained that the money he got from the government to build his house was spent just in financing the cost of transporting the material to the remote site where the house was going to be built. This was because he had to employ labourers to transport heavy materials such as cement, bricks, tools, etc., by foot up a steep path. Steep construction sites also need to be terraced—a prerequisite for obtaining subsidies—which is an expensive endeavour. For rubber plantations (but also for any other trees), the distance to the nearest road matters, especially when the plantation is being clear-cut. As a matter of fact, some farmers consider it uneconomical to clear-cut a plantation as paying agricultural labourers to transport the logs would be so expensive. Many have decided to make charcoal on the spot and sell it on the market—still at next-to-no-profit. Others have decided to cut the branches of the rubber trees to reduce shade and to plant other cash crops instead (in this case, pepper).

Farmers living in the hilly areas of Thalanadu also mention the problem of water availability, especially in the dry season (December/

January to May/June). The small streams in those areas dry up faster than the bigger ones in the lower areas. Water infrastructure, such as water pumps, is far away. In some cases, neighbours who have a well with sufficient water help out, but otherwise these households have to go long distances to fetch water.[21] Such households, therefore, are unable to plant certain crops (e.g., vanilla), since their cultivation is dependent on good water availability.

The fact that some households have moved, or are planning to move, nearer to or into centres of population demonstrates the importance of advantages such as access to services, for example, drinking water schemes, electricity, phones, schools and health centres. It has already been mentioned that it is easier to find daily wage employment when situated near the centres, where more potential employers live. However, if a family moves nearer to the centre, they are further from their agricultural plots and which is then harder to manage.

### 6.3.2 Type II: Income from own Rubber and (Part-time) Non-agricultural Activities

Marginal holdings of Type II have the lowest average number of rubber trees (51) and of total land (0.29 ha) of all the types (Table 6.6). They derive their main income from non-agricultural activities, often practised on a part-time basis. Thus, income from agriculture, and especially rubber, does not play a dominant role in the households' daily livelihoods. This type of holding represents only a small part ($n$ = 5) of all the interviewed holdings ($n$ = 40); the income strategy of Type I is more relevant to the study area as a whole.

*Part-time activities in non-agricultural sector*

In each of these households, one or more members are engaged in wage employment jobs. One man works as a full-time sales clerk in a medical shop in the nearby urban centre, leaving for his job after having tapped his 90 rubber trees. In one Muslim family, the husband has a regular part-time job in the local mosque, while his sons had left the parental home in search of work in the city where they have all found employment, regularly sending part of their salary home. This family was, however, planning to sell their house and land and live in Eratuppetta, a nearby town. In another family, the son works seasonally and part-time as a construction worker. Finally, on one holding, two brothers are self-employed in their own Ayurvedic medical centres; the wife of one of them regularly prepares Ayurvedic medicine to be sold in these centres.

In this type of holding, intra-household labour division and the role of women are important, as has already been mentioned. In some households, women do take off-farm jobs. In one holding, a woman works regularly in a local petty trade shop belonging to a self-help group of which she is a member. On another farm, a woman migrates seasonally to the neighbouring state of Tamil Nadu to work in a T-shirt factory (Box 6.3).

| Box 6.3: Sobana, Working in a T-Shirt Factory in Tamil Nadu |
|---|
| The income of Sobana's household is quite small. Since the family cannot cope on the single income from agriculture, Sobana worked as a housemaid in different families in the north of Kerala. However, in 2003, Sobana saw a job advertisement in a local newspaper by an agency looking for women ready to work in a T-shirt factory in Tirupur, the big textile-manufacturing centre in Tamil Nadu. Some months later, having paid Rs. 1,000 to the agency, Sobana left for Tiripur and started to work in the T-shirt factory. However, she had to return after only four weeks, much earlier than planned, because she had some problems with her leg. She could not stand for 14-15 hours a day, with only two daily breaks for breakfast and lunch (working conditions were as follows: 8.30 a.m. till around 10 p.m., 7 days a week, two breaks per day). However, the fact that the job is very tiring has an advantage: not many women are prepared to work there and therefore, she can go back any time to work there. The only other problem she mentions is that she does not like the food in Tamil Nadu. However, since it is a good opportunity to earn money (Rs. 620 a week including accommodation), she had decided to go back to Tiripur for four months in 2004. After that, she was planning a holiday to see her family. At the time of interview Sobana was looking for a place in a government-sponsored hostel and school for her daughter, failing which, her husband would look after the girl. |

*Relation to own rubber*

Since at least one person in each of these households is earning a non-agricultural income, the income from agriculture is less crucial. These holdings also have the advantage of being able to invest their time in non-agriculture wage-employments (described above). But most still tap their rubber trees and live off the income from rubber. However, since income from rubber trees is of secondary importance, their management is also of secondary importance, and this leads, in some cases, to low yields of rubber (e.g., due to lack of fertilizers). Interestingly, some interviewees mentioned that they were planning to buy more rubber in the future to complement their non-agricultural work. In fact, they said that rubber was very complementary to other (non-agricultural) jobs, since it requires less management than other crops (e.g., vegetables), but still gives regular yields. However, they were looking to buy a young rubber plantation, as the first two years of replanting are very time-consuming as well as risky (loss of trees). Some holdings derive a certain income from other cash crops, alongside rubber, and animal rearing. It was mentioned that the combination of tapping rubber (every second day, very early in the morning) and a full six-day off-farm job is very strenuous.

### *Gender and intra-household relations*

Looking at intra-household relations, a specific characteristic of these holdings is a strong collaboration between family members, which on smaller farms often means between the husband and his wife. For example, the husband taps the rubber early in the morning but soon after finishing, he leaves the house for his job in town. Therefore, his wife collects the latex and processes the rubber sheets. Women also pursue other agricultural activities, for example animal-rearing. Doing an external job would be difficult if there were no possibility of sharing the tasks in the rubber plantation.

Another interesting intra-household aspect is the differentiation in the patterns of expenditure. There was a clear distinction in some holdings in the use of the different incomes. For example, in the household of the Ayurvedic doctor, the income from his job was used for daily household expenses (food and others items, schooling, etc.). However, the income from the few rubber sheets was reserved strictly for emergencies. Thus, his wife stocks all the rubber sheets until an emergency (for example, buying medicine) should arise. At the time of the interview, she had stocked twenty-five rubber sheets.

*Geographical household pattern*

Houses in the holdings are located closer to the main road than in those in Type I holdings, for example. One hypothesis is that these holdings were able to afford more expensive plots near the road due to their more or less regular non-agricultural income.[22] Though the houses are simple and in the same style as the ones in holdings of Type I, their location gives them improved access to infrastructure such as transportation (public and private), phone, electricity and water.

### 6.3.3 Summary

Rubber farmers with holdings of Types I and II have to carve out a livelihood with minimal resources and a low average income. Their average land and rubber trees are insufficient to secure their livelihood and they are obliged to find other sources of income. Thus, most people look for wage-employment in agriculture and non-agriculture. A characteristic of all these holdings is their extreme dependence on their daily income from the different sources to make ends meet. Most of them in fact, have no savings, neither in cash nor in kind. This makes them highly vulnerable, especially when sudden events such as accidents or diseases occur. If these have long-term consequences (disability, medication), the household can be plunged into extreme poverty.

The most important difference between households of Types I and II is that the former depend strongly on agricultural income (both on their own farm and from wage income in agriculture). As a consequence, their income is much more dependent on seasonalities than is the case in Type II. For holdings of Type I, this implies a higher income risk and fluctuation.

The income pattern of these holdings also impacts on the way they relate to their own rubber plantation. For these holdings, rubber is an important crop due to its regular yield and its comparatively low requirements in terms of cultivation management. However, rubber is and remains for most of them a secondary income and its management is thus suboptimal compared to other types of holdings. In most marginal holdings, the mixed crops that are sometimes planted within a rubber plantation do play an important role for subsistence.

## 6.4 INCOME STRATEGIES OF SMALL HOLDINGS WITH MEDIUM INCOME

Type III holdings, that is, small holdings (1-2 ha) with a medium income (Rs. 50,000-150,000 per year), manage their rubber plantations as best they can, since they depend to a great extent on them for income. Thus, they specialize rather than diversify. However, the specialization of these holdings is especially interesting compared to the diversification of Types I and II described earlier. In fact, an understanding of the reasons for these holdings' specialization in rubber can give more clarity about the reasons why others diversify out of natural rubber production.

### 6.4.1 Type III: Income Based on Natural Rubber Production

#### *Role of Rubber Plantations*

Type III holders have an average of 1.40 ha of total land and 316 rubber trees (Table 6.6). The latex collected from these trees is generally sufficient for a household to secure its livelihood. Table 6.8 gives an overview of the average gross annual income of these holdings.

TABLE 6.8: AVERAGE GROSS INCOMES (IN RS. PER YEAR) OF TYPE III, PER SECTOR ($n = 9$)

| *Size category* | *in Rs. per year* | *in %* |
|---|---|---|
| Agriculture, on own farm | | |
| from rubber | 44,258 | 42% |
| from other crops | 18,120 | 17% |
| Agriculture, off-farm | | |
| from plantation on leased land | 22,222 | 21% |
| from wage employment (tapper, *kooli*) | 0 | 0% |
| Non-agriculture | | |
| from external employment | 6,667 | 6% |
| from self-employment | 14,886 | 14% |
| Total | 106,153 | 100% |

*Source:* Author's interviews.

Type III rubber holders are characterized by their knowledge as well as understanding and experience of the rubber plantation business; they also work to improve their plantations.[23] This interest is certainly linked to their dependence on the income from rubber. Further, some interviews suggested that there is never any question whether or not to replant rubber (once an old plantation is removed). Rubber is culturally strongly integrated in these holdings and would never be replaced by other crops. Thus, it seems clear that rubber will also be replanted in the next generation. This is also linked to the many advantages that rubber has compared to other crops such as coffee or pepper. The most important advantages mentioned were the regular yield of rubber, the fact that it does not require much care after the immature period (only manuring), and that its yield is not very dependent on the climate.

A characteristic of these rubber holders is that they almost exclusively use their own household labour to tap rubber. Wage labourers (*kooli*) are only hired for special seasonal work such as weeding and manuring, or during replanting work. The reason for their unwillingness or inability to employ external tappers is that their margin would then be too low; that is, if they employed external tappers, the income from rubber would not cover their household expenses. As one farmer said, 'If we would employ tappers, we would not be able to eat rice.' This is an important difference from holders of Type V who can afford to employ external tappers. Concerning the daily management of the rubber plantation, it is important to highlight the labour division within the households. Since external tappers are not employed, different family members, including women, take care of tapping, collecting the latex and making and drying the sheets.

### *Other agricultural income*

In many of these holdings, income from crops other than rubber plays a certain role. The focus is mainly on traditional local crops such as coffee, cocoa, coconut, arecanut, pepper and tapioca, as well as on some newer crops such as vanilla, planted next to the rubber plantations. One difference from marginal holdings is that in this case subsistence is not an important reason for planting these crops, although many do like to have basic food items (e.g., coffee, coconut, pepper) from their own gardens. One staple crop that is grown is tapioca, which is planted mostly in rocky areas where nothing else grows. Tapioca is used as a food crop in the household and the rest is sold on the local market. The decision to plant these cash crops is thus a reaction to short-term market fluctuations, and there seem to be regular switches from one crop to another. In times of low rubber prices, shade created by the rubber trees becomes an important limitation on growing these cash crops. Another constraint is the heavy workload resulting from cropping many different crops together. A further agricultural income mentioned by a few holdings is from milk and pork production. For one holding, the rearing of pigs was an important secondary income source (Box 6.4).

**Box 6.4: Shaji and His Interesting Pig-Rearing Business**

Shaji and his family like farming. They are trying to improve their income through new opportunities and strategies of cultivation. Some years back, they were looking for a way to use all the jackfruits that fell from their trees and went unused. Then they heard that pigs eat jackfruit and decided to start a pig-rearing business. Instead of fattening the pigs themselves, they decided to keep two sows and one boar and to specialize in rearing small piglets to be sold to other farmers for fattening. In optimal circumstances, a sow has 8-10 piglets 2-3 times a year. After 45 days, each piglet can be sold for around Rs. 800. For the family, it is an excellent business since they do not have to buy much fodder; most of it comes from their own farm, such as jackfruits and the waste from tapioca cultivation and processing. They are happy to be in this business; however, they mention that it requires much attention and care to breed and rear small piglets. But since they like activities on their farm, they do not mind.

The cultivation of some newer crops like vanilla is only possible because these households have available land and capital,[24] and enjoy a certain risk-bearing capacity for such investments. In fact, planting new crops is linked to plantation risks (unknown management practices due to the lack of long-term experience) as well as market risks (unclear market demand). Looking at the holdings of Types I and II, it is assumed that they are not able to plant these crops because they do not have the investment and risk-bearing capacities (section 6.3.1).

Household members of this Type III holdings neither engage in agricultural wage labour activities nor work as employed tappers in external rubber fields. This is a special characteristic of these holdings which differentiates them from the marginal holdings. Some mentioned that the social status of their family does not allow them to pursue these activities. However, in one case, one member of a household mentioned that he accepted such work when the rubber price was too low to sustain the household with it alone during the 'rubber crisis'. One farmer was also interested in leasing some external land in order to plant a banana plantation. However, due to the social status of his family, he was not allowed to do so. In fact, leasing land is only socially accepted if the leased land reaches a minimum size of around 3-4 acres (1.2-1.6 ha). This is because then he is accepted as a businessman investing in foreign land and employing labourers to do the plantation work. If the leased plot is smaller, the family members would have to do much of the work themselves.

*Starting non-agricultural jobs*

At the time of the interviews, some farmers of the younger generation were starting self-employed non-agricultural activities, though this is not a general characteristic of these holdings. Though their primary income was still derived from rubber, they used their spare time (mostly in the afternoons) to work in their newly started companies. In one case (Box 6.5), a farmer entered the rubber-sheet marketing business, which consists of buying sheets of second quality, washing them, smoking them and selling them as better quality sheets.

In a second case, a farmer collaborated with a colleague to start a company specializing in producing food snacks using dried beef. He was still in the testing phase, but the business idea seemed promising. Interestingly, both had started their business because of the desire and/or need to improve their respective incomes, and the need for other activity besides their rubber farm. In 2004, instead of using this income for household expenses, they reinvested it in their business.

There were two other cases of part-time wage-employment by owners of this type of holdings. One farmer worked as a part-time engineer for the state phone company (BSNL). However, he did not have civil servant status and thus did not get a good salary or any other advantages (pension, etc.).[25] His hope was that he would get a stable, long-term job. Another farmer was employed as a kitchen employee in the Indian Coffee House

**Box 6.5: From Farming to Business:**
**Thomas and the Start-up of his Business**

In 2002, Thomas decided that he wanted to invest his energy in some other activities besides managing his rubber plantation. Since he had spare time in the afternoons, he decided to use it for something that would also provide his family with some additional income. Since his family owned a smokehouse for smoking rubber sheets, he collected sheets of bad quality from a neighbouring rubber sheet shop, smoked and cleaned them, and sold them again at a better price. After seven months, he had already realized that this was quite a profitable business. However, unexpectedly, the owner of the shop that supplied Thomas the rubber sheets decided to smoke the sheets himself, and refused to sell them to Thomas. Thomas looked around for other shops which could provide the sheets to be smoked, but his cousin (also a rubber sheet-shop owner) told him that he ought to start his own shop and buy the sheets directly from the farmers. Thomas decided to do this. With his cousin's help and a bank loan, he started the business and opened a shop in Eratuppetta, the local business town. Two years later, he was earning one quarter of his total income from this work. He thinks that 'in some time' he will be earning more from this business than from rubber cultivation.

in a city around a hour and a half away from his home. However, in all the cases mentioned here, agricultural income remained higher than non-agricultural income.

### *Vulnerability and dependency on rubber price*

Holdings of Type III are clearly less vulnerable than those of Types I and II. In fact, they have some risk-bearing capacity to overcome difficult situations. For less significant events, they have small stocks of crops which can be rapidly sold. In the event of larger (sudden or planned) expenses, they can cut down wild trees or, if necessary, sell land or the wood from an old rubber plantation which needs to be replanted. Another option is to get a loan from a bank against the deposit of land documents. All these strategies are used to pay the dowry when a daughter gets married.

One potential point of vulnerability is linked to the development of the natural rubber price. Rubber price plays a crucial role for most of these holdings, as rubber represents, on average, the most important source of income and, thus, holdings of this type are the most sensitive to rubber price changes. The extreme fluctuations in the price of rubber during the period 1994-2004 had a deep impact on these holdings. As one interviewee mentioned, the members of his household had to 'adapt their living style to the income from rubber'. On the other hand, specializing in natural rubber means holders can improve their management practices, potentially leading to higher yields.

*Geographical pattern*

A characteristic of these holdings is that all but one are located in Maravikkallu. Most of the holders live in traditional, often well-renovated houses. These houses are all located near the main road running across the Panchayat and they have good access to this road. The majority of the families have been living in Maravikkallu for more than one generation; in most cases it was the grandfather who built the ancestral house.

### 6.4.2 Summary

Rubber farmers of Type III mainly specialize in rubber. They cultivate rubber following the recommendations of the Rubber Board and most of them are expert planters who prefer rubber on their plots. Natural rubber is also culturally integrated into their daily lives, and they will always replant rubber. Tapping is done almost exclusively by household labour because the size of the plantation does not make employing tappers economically viable. Some of them do plant other cash crops depending on which will bring the greatest income. Some have also successfully integrated animal-rearing into their farms.

The social status of household members of this type does not allow them to engage in wage activities in the agricultural sector. This characteristic differentiates them from the marginal holdings of Types I and II. Instead, some younger farmers have ventured into new non-agricultural activities in the services sector.

These farmers are less vulnerable than the marginal ones. For bigger and unexpected expenses, they can afford to sell some of their lumber trees, or cut and sell the trees of an old plantation, or, at worst, sell some of their land. Their most important point of vulnerability is their dependence on the rubber price and this was felt during the 'rubber crisis' (section 6.6.2).

## 6.5 INCOME STRATEGIES OF SMALL TO MEDIUM HOLDINGS WITH GOOD TO VERY GOOD INCOME

This section documents the income strategy of holders with good to very good income (> Rs. 150,000 per year). First it details the strategies of small to semi-medium holders who however, due to a full-time and well-paid wage-employment, bring in a good to very good salary (Type IV). Next, it gives the details of semi-medium to medium holders which base their income on agriculture, and have, as well, a good income due to their large land holdings (Type V).

### 6.5.1 TYPE IV: INCOME FROM A FULL-TIME WAGE EMPLOYMENT AND OWN RUBBER

*Full-time wage or self-employment*

The biggest source of income for these holdings is not derived from agriculture, but from the non-agricultural sector (Table 6.4). In most households, at least one member is in government employment, while in two cases household members were self-employed in the service sector.

Of those in government[26] employment, some had very good positions (with a good income), while others had an income similar to what they could potentially earn from their agricultural land. However, the advantage of a civil service job *vis-à-vis* agricultural income lies in the good government employment conditions with a regular salary, an insurance scheme, a pension scheme, etc. Thus, government employment does not present the same risks as agricultural production which is under threat from changing prices, irregular climatic conditions, and seasonalities. Of the self-employed, one person was an Ayurvedic doctor with a well-known medical practice in a town nearby. The second had a very diverse income pattern, owning a tea shop (a local restaurant) and a taxi company, and driving the local school bus (Box 6.6). Finally, there was one case of a family getting irregular remittances from family members, as well as some rent from the leasing out own land for slaughter-tapping.

*Rubber management and income from agriculture*

Holdings of Type IV have an average of 1.59 ha of total land and 366 rubber trees (Table 6.6). Theoretically, income from these rubber trees should be sufficient to cover the average basic household costs at least. However, since these holdings have profitable non-agricultural activities (see above), it is not a necessity for them to maximize income from rubber.[27] This is also demonstrated by the fact that, in most of these households, rubber tapping is done by externally employed tappers and not by household members. However, this greatly diminishes the income from rubber, as the interviewees themselves often stated. The main reason for employing tappers is a lack of time, the size and the scattered locations of the plantations, but in some cases also a lack of tapping skills within the households was the reason. This lack of time is of course linked to

**Box 6.6: Joseph and his Multiple Non-Agricultural Income Sources**

Joseph is 35 years old and lives with his wife, their two children and his mother in a house in Thalanadu. They own two hectares of land from which they have some income. However, since Thomas never learnt how to tap rubber, he is not working on his plantations. Thomas' main work is to run a tea shop in a town nearby. Eight years (before the interview) back, his father rented the tea shop and after his father's death, Thomas took over. He employs his brother and two other part-time staff. This business is not very profitable but since he cannot do much on the farm, he likes to work there. Also, his family needs some additional income. However, he does not know how long he will be able to continue this business since the owner of the building is not sure whether he wants to renew the contract. This insecurity—but also the fact that Joseph is still looking for a more profitable business—convinced him to start another business. Four years back, he and his brother bought a new jeep with the help of a bank loan and private moneylender and started their taxi enterprise. Nowadays, whenever a customer needs their taxi, Joseph or his brother will drive. They do this way because they cannot afford to hire a driver. Since they did not manage to repay the loan to the bank, Joseph decided to take on another job. Thus, every morning, he drives the official bus of the local school. His family manages to make a living with these three incomes, complemented by some income from agriculture. Though they are not poor, they still live with regular moments of insecurity because they do not have the regular and long-term income of a governmental employee.

their wage- or self-employment outside agriculture. Interestingly, many of these holdings mentioned that they have difficulties managing the employed tappers. In one case, the family had tried to work with around six to seven different tappers before finding the one they were working with at the time of this research. But still they were unhappy because they suspected that the tapper was not delivering all the rubber sheets to the farmhouse. They said that the problem was their lack of time to supervise the work of the tapper on the distant plot. Following this, they planned in the mid-term to replant the plot with young trees and sell it for a good price in order to buy a different plot closer to their house.

Besides rubber, some holdings gain an income from cultivating mixed crops including lumber trees. Since the income from agriculture is not critical, the choice of the cropping pattern seems to be based also on factors other than merely economical ones. For example, one farmer said that he plans to stop the cultivation of rubber and focus on the production of mixed crops once he retired from his civil service job. His dream is to develop a homestead garden for his family. Another farmer has already cut down some rubber trees to be able to plant mixed crops along the same reason; however, he has kept some rubber trees because of their greater drought resistance.

*Geographical pattern of holdings*

Holdings of Type IV are located near main roads and almost all have electricity and telephone connections. In one case, at the time of the interview a farmer was still living in a hilly place far from the road but some months earlier he had bought a new housing plot including a contiguous rubber plantation, close to the main road. The family planned to build a house and move there in the mid-term in order to have better access to school facilities for their children, as well as other facilities such as transportation, electricity and phones. However, the farmer mentioned that the distance to the other rubber plots belonging to the family could become a constraint. This example shows that, whenever possible, families prefer to move to near the main road (or near centres) rather than live in hilly areas with less access to services. But most families are not as eager to move to bigger towns or cities. In fact, some families who would have been better served by living near their government offices in big cities preferred to commute every day to their places forward. They like the village life.

### *Long-term planning*

Some of the comments on the long-term farm planning—and, related to this, to the future employment of their children—are interesting in the case of these holdings. As it happens, one holder did not know how to tap because his father (a pensioned tapper) had encouraged him to study and to engage in other business rather than become a tapper. On two other holdings, the son and daughters were not interested in farming and had trained for other jobs. The consequences relating to rubber were twofold: some parents kept rubber as an insurance, just in case their children did not get a well-paid job. On the other hand, if it became clear that rubber would become rather difficult to manage and their children would seek jobs outside the village, they planned to reduce the area under rubber.

### 6.5.2 Type V: Income from Own Rubber and Other Cash Crops Planted on Leased Land

Holdings of semi-medium and medium size have, on average, 3.41 ha of total land and 971 rubber trees (Table 6.6); they are therefore the largest holdings in the area. In these households, the income from rubber and from other cash crops on the farm covers all the household expenses. Still, many farmers on these holdings venture into the seasonal business of planting cash crops on leased land.

### *Rubber income and management*

Type V farmers generally specialize in rubber cultivation. Most of their rubber trees are planted in accordance with the recommendations of the Rubber Board. They follow most technical advice from the Rubber Board because most planters trust their scientifically tested management practices. This means, for example, that only the maximum permitted number of non-rubber trees are found within the plantations, as this is a requirement to get replanting subsidies (section 7.2.1). Also, these farmers are in regular touch with the village field officer of the Rubber Board, and they also know staff at the headquarters of the Rubber Board with whom they sometimes have exchanges about specific issues.

The large size of their rubber plantations means they have to employ external tappers, though many of these farmers still tap part of their plantations themselves. This, however, depend on their availability and the distance to their rubber plantations. One constraint mentioned is the difficulty in employing skilled tappers. In one case, this led to a situation where one tapper had to postpone the tapping of his newly planted rubber plot by three years, starting only when the rubber trees were ten years old (instead of the ideal seven).

*Income from other cash crop on own land*

Many of these farmers also plant other (sometimes mixed) cash crops and lumber trees on their own farm. Coconut, coffee, tapioca and banana—all planted in plots close to the rubber plantation—play the most important role. In a few cases, income from these crops was quite important and as high as that from rubber. It is often out of personal interest that farmers diversify into mixed cropping and not due to subsistence needs, as is the case in the holdings of Type I. For example, in one case, an elderly farmer cut down the rubber trees around his house because he was feeling 'suffocated' by them. Instead, he planted young coconuts because he felt there was nothing nicer than 'to see the first coconut growing on a young tree'. Occasionally, these farmers do test new crops, cropping systems (e.g., organic cultivation) and technologies (e.g., vermi-composting, biogas, drip-irrigation). At the time of the field survey, many had started to plant vanilla; along with the high market price for vanilla, the subsidy scheme from the Panchayat provided an added incentive (section 7.2.3).

*Cash crop plantation on leased land*

One of the main strategies of these rubber holders is to invest in seasonal plantations on land leased from other farmers, a self-employed business. The most common cash crop is banana or plantain, followed by pineapple. In one case, a farmer leased land to start a rubber nursery with rubber plant seedlings. Although leasing land is officially prohibited in Kerala,[28] it has become a common practice of larger farmers in this Panchayat. One important reason farmers gave for leasing land was that they wanted 'to be active also in the afternoon'.

The leasing agreements are very diverse, some being completely non-monetary in their terms (Box 6.7).

Many farmers mention the high investments and initial capital needed and the consequent risk linked to leasing activities, especially if they have had to take a bank loan to pay the lease. More risks are connected with the short-term development in the crops' price, but also because the management of the plantation is heavily dependent on the timely availability of good and inexpensive labour in the area surrounding the leased plot. Furthermore, good road access and water availability is crucial. However, the high risk normally leads to a good return on investment. But, why can these farmers (Type V) manage to lease land while others focusing on agriculture (e.g., Type I or III) cannot? The hypothesis is that leasing activities (or other risky businesses) can only be undertaken because Type V holders have a good income from rubber to backstop the risks.[29]

As mentioned above, one limiting factor is the availability of labour.

### Box 6.7: George and his Two Leasing Agreements: Pineapple and Banana

George's family owns around 1,180 rubber trees, of which he regularly taps 600. Besides his involvement in the rubber plantation, George has entered into two leasing agreements. In the first agreement, he leased a six acre replanted rubber plot belonging to another farmer. In that area, George had been cultivating pineapple for three years at the time of the field survey. In exchange for the use of the land, George—who is well-known for his skills in replanting rubber—has to plant rubber seedlings and take care of the small rubber trees during their first three years. Since this is a very critical stage of replanting, the landowner is happy that it is George taking care of them. And George expects to get a good income from the pineapple crop. Thus, in this agreement, no monetary exchange is involved.

In the second three-year agreement with the same landowner, George planted 2,000 pits of banana in a plot of five acres. This was the third consecutive year that he was been allowed to plant banana on this plot. In exchange, he also planted young mahogany and teak trees on it for the landowner. For George, it is a good business since the land is very fertile. Up to the time of the interview no monetary transaction involved. However, from the following year onward, he would have to pay had been a leasing fee since the three-year agreement would have expired. He was thinking about finding a new arrangement since he had also planted some tapioca on that land.

According to some farmers, the availability of enough labourers for a reasonable price is one of the key reasons for choosing a region in which to lease land. In fact, closeness to a village or a hamlet where many agricultural labourers (*kooli*) live is an asset. However, even employing this tactic does not solve the huge problem of the lack of labour for these farmers and the whole agricultural sector in Kerala.[30] In the interview it was frequently mentioned that growing cash crops on leased land is a physically and emotionally arduous job which requires the farmer to be in good health. This is the main reason some older farmers abstain from getting involved beyond the management of their own big rubber plantations.

### *Non-agricultural income*

A characteristic of these holders is that they only pursue a few activities in the non-agricultural sector. The main reason for this is their full involvement in the agricultural sector as described above. There were, however, two cases of farmers involved in the non-agricultural sector. One farmer was renovating his traditional house and transforming it into a guesthouse in order to start an agri-tourism activity. Another farmer had constructed a rubber sheet-smoking house, with the intention of buying second-quality sheets from neighbouring farmers and transforming them (through a process involving cleaning and smoking of the sheets) into first quality sheets (see example in Box 6.5). Finally, one old farmer, all of whose children except one live and work in the US, had the option of requesting remittances from them if needed. However, due to a good flow of income from rubber, this was seldom the case, and then only for 'emergencies'.

### *Geographical pattern and intra-household income management*

A feature these holdings have in common is their location. The holders are all in Maravikkallu and live in ancestral houses with direct access to roads (expect one) and to infrastructure for water, phone and electricity. Also, public transportation and schools are nearby. Furthermore, their rubber plantations are situated in the best, flattest locations of the ward. Finally, it is important to mention that many of these families have a common family background. Out of the eight farmers selected (at random) for interviews, six were related to each other (brothers and sisters, or cousins).

Looking at intra-household relations, the specific income-sharing within the household is of interest. It was mentioned that income derived from natural rubber and from other crops on the farms was used to pay all household expenses. On the other hand, things were less clear concerning the income derived from off-farm activities such as leasing of external land. In fact, this income was often managed by the person who was involved (both monetarily and in terms of labour) in the leasing agreement. It did not emerge clearly from the interviews what the profits from activities on leased land are used for, but some interviewees specifically mentioned that they are not used for daily household expenses. This is an interesting intra-household aspect which, however, requires more detailed analysis.

#### *Long-term planning*

Many of these holders' children are getting (or have already had) a good education, often in a private college. Consequently, people from this new generation are more interested in finding other, less risky jobs than getting involved in the agricultural sector. As a result, many young adults were either off working in other Indian states or, occasionally, abroad (in the UK or the US). However, exceptions do exist: a young farmer who had studied computer engineering in Bangalore decided—after having worked some years in the IT business—to take over his parents' farm. He invested a lot of money in renovating the house and planned to start an eco-farm tourism business.

On the other hand, parents also make sure that the son (or daughter) taking over the farm can make a living from the income; thus, a common strategy of some farmers is to enlarge the farm by buying new plots whenever possible.

### 6.5.3 Summary

Types IV and V farmers possess large plantations and a number of rubber trees, that normally allow them to make a comfortable living. However, their livelihood strategies are different. Type IV holders do not base their main livelihood on rubber but on civil service jobs or on profitable self-employment (e.g., as a medical doctor). The advantage, especially of a job in the civil service, is that it is almost without risk compared to agriculture. Due to the lack of time to manage their rubber plots and supervise the labourers (tappers), their plantations are sub-optimally managed. Thus, the income from rubber is lower than it could be. Type V holders, on the other hand, base their livelihoods on their rubber plantations. They manage their farms with the latest available knowledge and technology. Almost all employ external tappers—in some cases Type I farmers—since they could never tap all the plantations using household members due to their size. Sometimes mixed crops (e.g., vanilla) are added to the farm and these farmers like to test newer crops, technologies and cropping patterns. The second-most important source of income for these holdings comes from planting cash crops on externally leased land. Leasing conditions vary according to the agreement. Leasing, a high-risk activity, can be done because the income from rubber can potentially cover (at least part of) the risk involved. But if the conditions for leasing are good, the profit can be very high.

Both types of holdings have large, traditional, often renovated farmhouses in common. Finally, an interesting finding specific to these holdings is linked to the future of their children (or the next generation). The relative wealth of these families means that their children can go to privately run schools, colleges and, if they are interested, universities. Consequently, many of them are not interested in working in the agricultural sector anymore and their parents do not push them to do so. This has consequences and risks for the long-term planning of these farms. However, investing in land is considered one of the best strategies as the land can always be managed by the family, or if not, sold.

## 6.6 IMPACT OF AND REACTION TO THE 'RUBBER CRISIS' (1997-8 TO 2001-2)

Interviewees spoke about the general impact of the 'rubber crisis'. They often mentioned that the price of consumer goods and services (including the salary of labourers and tappers) went up when the economy faced a period of high natural rubber prices around 1996-7 (Fig. 4.7). However, these prices remained at the same level during the 'rubber crisis'. As an example, a tapper's salary went up to about Rs. 25 per hundred tapped trees, and it remained at this level during the crisis. Rubber holders frequently cited this problem.

The biggest problem the farmers faced was the sudden change from an unexpectedly high price of natural rubber to a very low price. In fact, when the rubber price was high, many rubber households started to spend far more money on (luxury) consumer goods as well as on land and house constructions. In many cases, they took expensive loans to do so. Many agreed that the lifestyle of most farmers was very luxurious during that time; people spent lavishly. However, suddenly the price of natural rubber inflected and fell to a very low level. All farmers who had taken loans found themselves in a difficult situation, especially those whose income was highly dependent on rubber. As some said, the whole local economy—including shops and the service sector—was affected.

However, the 'rubber crisis' had some positive consequences for those who could still afford to buy land: the low price of natural rubber also drove the price of rubber land and rubber replanting down. The fact that some farmers needed money and had to sell their land quickly was advantageous for others who were able to buy more land at an attractive price, most of it already planted with rubber.

### 6.6.1 Impact on and Reaction of Marginal Holdings of Types I and II

Interviews showed that the low price of rubber did not affect the farmers' decision to plant or have planted rubber; nobody thought that they would have to cut down then rubber plantations because of the low prices. The reasons given for this are mainly the advantages of rubber, such as the low management requirements of a mature plantation as well as its regular yield and income-generating ability. Thus, rubber was tapped normally and according to the normal rhythm; because of the low tapping costs (rubber mainly tapped with family labour), it was better to have a small income from rubber than to have no income at all. However, during the crisis, farmers dedicated less time to their rubber plantations, for example, they stopped spreading manure.

Farmers generally mentioned that the overall cropping pattern on their farms had not changed much over the last eight years, i.e. from the start of the crisis around 1997 until 2004, and, despite the low rubber price, the share of rubber in the crop mixture remained constant.

Although rubber possesses undoubted advantages, farmers also seemed to be lacking viable alternatives to rubber; in an area where 'rubber is growing all around', there is almost no space and light for other crops. Interestingly, some farmers did not only maintain their rubber trees but also replanted during the times of low prices, sticking to their normal replanting agenda. Again, the advantages of natural rubber such as its drought resistance are instrumental in the choice to keep on the crop. Finally, farmers said they were aware of the price fluctuations; during the low price times they expected rubber prices to start rising again after three to four years, as many crops show this type of price cycle.

As seen above, many of these farmers are employed as tappers. Since the salary of these tappers did not go down during the crisis, their wage income stayed mostly unchanged, and was only affected by the decision of some farmers that tapping was no longer viable. Yet, some farmers employing tappers decided to tap even during the crisis, mindful of the employment situation of their tapper.[31] Marginal holdings (especially of Type I) were thus especially affected by the lower prices for their rubber. However, since only part of their income is dependent on their own rubber, they were proportionally less affected than the ones who are dependent exclusively on rubber (such as Type III farmer).

Two farmers had to find a new job during the crisis. One farmer who had stopped working as an external tapper in 1992 due to eye problems had to restart some of his tapping activities. He also tried to find employment as an agricultural labourer (nowadays he is more passive: he will only take on work if somebody calls him, but will not seek it actively). In 2002 (in the midst of the crisis), another tapper decided to look for another job. He worked for eight months as a security guard in a town a half-day's journey from his home. His aim was to earn more money. However, he did not get the salary he was promised and was also spending much of it on his food. In the meantime, his wife could not manage to cope with all the work on the farm on her own and returned to her parents' house with their sons. Because of all these problems, her husband gave up his job and returned to the village to start working as a tapper again. Today, he says he is happy with that decision: although he does not earn more as a tapper, he does feel 'mentally free' and is with his family again.

Marginal rubber holders mentioned that, during the crisis, it was an advantage to have mixed crops growing on their land. These enabled them to balance the price fluctuations of crops with different prices. For example, the price of pepper was very high during the crisis in the rubber sector (Fig. 6.2).

Summarizing the changes in the agricultural income of holdings of Types I and II, it can be said that rubber income from their own farm

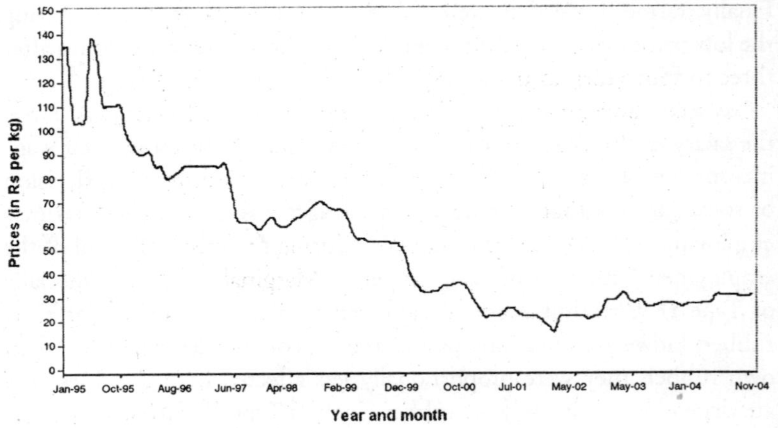


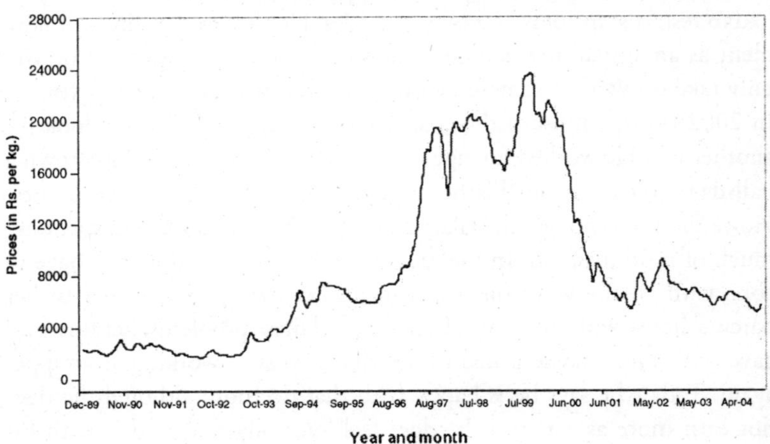


*Source:* R. Ramakumar et al. (2005).

Fig. 6.2: Price development of coffee (1995-2004) and pepper (1990-2004) in Kerala.

declined, but since the share of rubber is quite small (compared for example to holdings Type III), it did not have such a drastic impact on these farmers. Furthermore, the income from other mixed crops during these years was still good as these crops had a constant or higher average price. Agricultural income from tapping did not change much since the total number of tapping days and wages remained unchanged. But because employment for agricultural work (*kooli*) was reduced to the minimum necessary on big holdings, it was more difficult to come by these irregular jobs. The income of labourers from *kooli* work is shown to be very dependent on the specialization of the labourer. For example, labourers specializing in felling rubber plantations and loading/unloading wood enjoyed a good employment situation during the low price period.[32]

### 6.6.2 Impact on and Reaction of Small Holdings of Type III

Type III holdings are very dependent on the income from rubber and it was expected that they would have suffered the most from the low rubber prices. This was in fact the case, unless a family was able to balance it with income from other crops (pepper, coconut and tapioca were specifically mentioned, as they all had relatively good prices at that time). However, with regard to rubber, these farmers confirmed that it remained a very attractive crop in spite of the crisis. It appears that they did not modify their replanting plans despite the lower prices: on the contrary, due to the low price of rubber, (re)planting of rubber was also cheaper at that time. One clear advantage of these holdings was their exclusive use of household labour to tap their rubber trees. For many, this was the only way they could tap without incurring losses. Most could not have afforded outside labour.

An important constraint for some Type III farmers was that they had decided to fell lumber trees and some other fruit trees (coconuts) before the time of the crisis; in fact, they had to do so in order to receive the replanting subsidies from the Rubber Board. During the low price time, they therefore did not have lumber trees to cut and sell trees or the fruit from other trees. This is further analysed in section 7.2.1.

One farmer's strategy of focusing strongly on the income from his farm was of interest. During the 'rubber crisis', when the price of pepper reached a record peak; he cut the branches of his young rubber trees and used their trunks as support for the climbing pepper plants. During that time, he did not harvest latex. His plan was to farm pepper until the price of rubber rose above that of pepper. When this happened, he put more manure around the rubber trees to allow them to grow back their branches. His thinking was that this way of doing things would not affect the rubber trees; it would simply slow down their growth for some time. However, other farmers who did not follow these practices said that the ones who did were regretting their actions (no reasons were given).

Affected by the lower income from natural rubber, many households had to curtail their expenses. In practical terms, this meant that some of them ate less 'non-vegetarian'[33] food, reduced travel to a minimum and also cut down on telephone expenses. For many, the greatest difficulty was to make the transition from a period of high rubber prices, when they spent a great deal on consumer goods (refrigerators, etc.) to enhance their standard of living, to a period of low rubber prices when these expenses had to be reduced to a minimum. Some compensated for these losses in income with bank loans—especially 'gold loans'[34] —others ran up debts in the local grocery shops.[35] Some farmers had to sell part of their land to be able to feed their families. For others, the long period of low prices was a compelling reason to seek temporary employment as tappers or to start a new activity in the non-agricultural sector (Box 6.5).

Finally, as one interviewee said, the main lesson from this period was that families have to adapt their expenses and way of living to the present income (and thus not to their expectations of future incomes). The most affected families were those who had to repay loans during this period.

### 6.6.3 Impact on and Reaction of Small and Medium Holdings of Types IV and V

Types IV and V farmers have plantations of a size which permits them to gain a good income from it. This income was still sufficient to sustain the household during the low rubber price period, although expenses had to be reduced even on these holdings. However, the fact that Type IV holdings get most of their income from outside agriculture means that the income from rubber is less crucial. This fact was instrumental in some farmers' decision to plant other crops and diversify their cropping pattern at the time they were replanting rubber. Still, the good drought resistance of rubber was also a reason for them to retain a decent amount of rubber. Also, since land prices were low during this period, some bigger farmers of Types IV and V did buy more land, often including rubber plantations. In some cases they also got loans from banks to buy the land.

During the period of low rubber prices, some Type IV and V planters still made a profit from their rubber, despite having to bear the costs of employing tappers. For others, the income from rubber only covered the farmers' expenses. In these cases there were different strategies: some stopped tapping while others said that they still wanted to tap to keep their tapper employed, although they would make next to no profit from tapping.

Holdings depending on a member being self-employed in the non-agricultural sector (e.g., the case of the Ayurvedic doctor) were indirectly affected by low rubber prices. In fact, the doctor stated that he had fewer patients at the time, and that many of them did not pay—or paid only later—for their treatment. Another farmer who owns a tea shop confirmed that his turnover was sharply reduced during those years. This indicates that the whole local economy suffered from the low price of rubber. However, the tea shop owner said that even in present times (2004) people avoid spending too much money in his shop though the rubber price is higher. He thinks that people are now reluctant to spend money because they expect rubber prices to fall again.

To summarize, planters who had put themselves in a difficult situation (through buying land, constructing a house or buying luxury consumer goods) felt the impact most because they had speculated that the high prices would continue. Those who did not were not heavily affected; by reducing their expenses to a minimum they could live on their lower incomes. However, some were also indirectly affected because their non-agricultural businesses suffered from the local economy going through a slack period. Finally, the experience of the low rubber price period is still present in people's minds and still affects their spending behaviour.

## 6.7 TOWARDS AN ANALYSIS OF THE ROLE OF DIVERSITY AND DIVERSIFICATION

This section starts with an analysis of the current status of diversity, as well as of the process of diversification in the different types of rubber holdings (6.7.1). It continues with the analysis of the driving forces leading to households having diversified incomes and to the diversification processes (6.7.2). Next, the outcomes and impact, advantages and disadvantages of diversifying and of having diversified are analysed by type of holding (6.7.3). Finally (6.7.4), the findings are summarized and conceptualized.

### 6.7.1 Diversity and Diversification in Natural Rubber Holdings

Before embarking on the analysis, it is important to remind ourselves of the important role of natural rubber for most of the holdings. All the advantages of rubber cultivation described in the previous sections make rubber a crop which few are ready to do without on their holdings. However, for diverse reasons, households need or want to engage in other activities, in the agricultural and non-agricultural sectors. This section describes the role of these 'other' activities in rubber holdings without neglecting the role that rubber cultivation plays income-wise, but also socioculturally, for households.

Table 6.9 summarizes the diversity and diversification of activities and incomes on each of the holding types and assesses their importance. The analysis differentiates between *diversity*, i.e. the diverse status of the holding during the time of field interviews, and *diversification*, which refers to the process of change from a portfolio of activities to another one within a specific time frame (see theoretical debates, section 2.2.2). In this analysis the time frame looked at is the period from 1995-6 to 2003-4.

As Table 6.9 suggests, diversity is high in Types I, II and IV. Holdings of Types I and II have been more diversified for longer as they cannot make a living with only one type of income. This is why the diversification process was only rated 'low to medium' on these holdings. In Type IV holdings, there is a similar situation. As most non-agricultural activities were pursued by these holdings even before the 'rubber crisis', which started in 1997-8, the process of diversification is not recent. Thus, diversification is assessed as being 'low to medium'.

Most of the income on Type III holdings derives from rubber; but many also follow some other activities with a smaller income. Their income thus shows 'low to medium' diversity (Table 6.9). However, the process of diversification is evaluated to be in the "medium" range. This

TABLE 6.9: ACTIVITY AND INCOME DIVERSITY, AND DIVERSIFICATION

| | *Diversity (status)* | *Diversification (process, 1995/6 to 2003/4)* |
|---|---|---|
| Type I | *High*: mainly within agriculture, including agricultural employment | *Low to medium*: especially towards agriculture niches (e.g., livestock) and (part-time) non-agricultural employment |
| Type II | *High*: in non-agricultural sector | *Low to medium*: many already with fixed opportunity in non-agriculture |
| Type III | *Low to medium*: mainly rubber, but some few activities in other sectors | *Medium*: some manage to start non-farm business activities |
| Type IV | *High*: in non-agricultural sector | *Low to medium*: most have followed same activity for a long time |
| Type V | *Medium to high*: in agricultural sector | *Medium*: some start leasing activities |

*Source:* Author's interviews.

is so because some younger planters have recently started to venture into new off-farm business activities (e.g., production of snacks) with no direct link to their farm.

The biggest holdings (Type V) specialize in rubber cultivation, which provides them with a good income. The cultivation of other cash crops also plays a certain role. Thus, the diversity of these farmers is 'medium to high' (Table 6.9); the main difference as compared to the Type IV holdings is that the diversified income is almost exclusively derived from the agricultural sector. The process of diversification is assessed as being 'medium', since many farmers did start these leasing activities during the last years.

### 6.7.2 Driving Forces Leading to Diversity and Diversification

It is important to analyse the driving forces (or determinants) of diversity and diversification as they relate to the analysis of the role of the local institutional setting (chapter 7).

The analysis of the driving forces of the marginal holdings (Type I) shows the importance and role of vulnerability. This vulnerability is due to factors both internal and external to the holding. The *exposure* to threats (e.g., droughts or the death of a family member) is not much greater than on other holdings (with the possible exception of the risky location of certain houses of these holdings). However, more important is the high *sensitivity* of these holdings to these threats if they were to occur. Due to their low income, owning to the fact that it is impossible for them to save money and their consequent low risk-bearing capacities, these holdings are very sensitive. Although they have certain opportunities to access informal credit institutions (section 7.2.4), these options are limited both in size and in time. Thus, vulnerability determines the importance of a diverse portfolio of activities and incomes (in cash and in kind). Type I holdings *are pushed* into diversifying in order to cope. An important strategy to reduce vulnerability is ensuring that more than one household member is in wage-employment. On these holdings, women are also in off-farm wage-employment; in all other types of holdings they work exclusively in the household and on the farm. Since they bring home an important part of the holding's income, the risk of income failure is spread among more members of the household. Thus, if one member cannot work anymore, there is at least always somebody else bringing home a wage. This reduces the sensitivity of the holdings, which is ultimately a positive outcome of diversification. Another driving force is the geographical remoteness of these holdings. The difficult access to markets to sell and buy products makes on-farm agricultural diversification necessary for their subsistence. Remoteness is also a constraint to finding employment in the non-farm sector, since opportunities are often far away, and thus work as a tapper and/or agricultural labourer in the locality is favoured. In some cases, geographical remoteness and the lack of road access are instrumental in defining the cropping pattern (e.g., the case of an old rubber plantation which could not be clear-cut because it is too expensive to bring the wood out of the plot), leaving no room for agricultural diversification.

Type II households, as those in Type I, are very dependent on their daily income and are unable to save money. This dependency is the major driving force pushing households to get their income from several different sources, mainly off-farm and non-agricultural employment. However, maintaining this diversified status is only made possible due to the crucial role of the household members who stay at home, often the women. Without their support in farming activities, especially in the rubber plots, household members could not engage in both rubber plantation and external activities and would have to choose between them. This was the case in one family where all members, especially the younger generation, are in off-farm jobs; thus, they decided to sell their farm (including the house) and rubber plantation and move to the nearby town.

In the third type of holdings (Type III), families primarily focus on planting rubber since they own plantations sufficiently large to sustain their households. Most holders like managing and working on rubber plantations and this is a driving force behind farmers focusing on rubber. Their 'strategy' of basing their livelihood on rubber means that they have to work on the plantation themselves and cannot employ labourers; they therefore have less time for other activities. However, recently, some younger farmers have decided to start new business activities (section 6.4.1). Though still in the minority, their reason for *diversifying* is worth an analysis. The main driving force they mentioned is the availability of spare time when they have finished tapping their own rubber plantations, and their wish to do something else besides managing rubber. For some, the experience of the 'rubber crisis' (and the fact that the social status of their family prevents them from doing agricultural labour work, see section 7.2.6) was another important determinant that pushed them to improve their income by adding a new activity. However, these holders confirmed that they would have found it difficult to start a new activity had it not been for the help of their friends, as well as of public and private institutions (e.g., banks, see section 7.2.4). It is also important to mention the assistance and support of the remaining household members to allow one of their own to start new off-farm activities.

The driving force of Type IV holdings for their members working in civil service jobs is that they get very good employment conditions (including protection against dismissal) and access to a solid and long-term welfare scheme (pension, insurance, etc.). These are the main reasons, besides the good and regular income, behind these holders' desire to find government employment. As some interviewees mentioned, a job as a civil servant also reduces dependency on agriculture. Since these holdings have enough rubber trees to make a living, their diverse income is due more to opportunities taken than to a crucial need to find other sources of income.

As discussed earlier, many (but not all) Type V holdings are quite diversified within agriculture, with households planting other cash crops on their own land and also on land they have leased. The driving force behind planting crops other than rubber on their own farm is the wish to have a mixed crop pattern, more out of a personal interest than for any pressing need. There are dual driving forces for leasing land. On the one hand, many holders wish to get involved in other activities besides rubber since they have sufficient spare time, especially in the afternoons. Being in the 'leasing business' is an indicator of a certain social status, but these farmers also have the intention of improving their income with these activities. Leasing land is a risky activity which can, depending on many factors, be quite remunerative or involve losses. An important determinant in leasing land is the risk-bearing capacity of the farmers involved and their leasing activity is backstopped by the income from rubber. These holdings also often have enough financial capital to be able to undertake the high initial investments necessary. If they do not have this money, they can have access to bank loans using their land as collateral; this has happened in some cases. Finally, their social network plays a crucial role for them being able to find suitable land for lease, since the quality of the land and the site are decisive for the lease to work out successfully.

In conclusion, it is important to restate that, in general, the 'rubber crisis' was not, as one would have expected, a major—or even important—reason for starting or increasing diversification—and this is true for all types of holdings. Only selected households whose income is very dependent on rubber (Types I and III) actually initiated new activities during the rubber crisis. Rather than striving to improve their income, most holdings reduced their expenses, or, in the worst case, sold land or felled rubber trees to survive the crisis.

### 6.7.3 Impacts and Outcomes of Diversity and Diversification

What are the consequences of the diversified activities and of the diversification of activities shown in the different types of holdings? The combination of activities of holdings of Type I described above (section 6.7.1), i.e. agriculture (mixed crops), agricultural wage labour, and some part-time employment in non-agriculture, have the following outcomes and impacts. Farmers' diversified cropping activities on their own land have both advantageous and disadvantageous consequences. The advantages are the availability of subsistence food for the household, a regular income due to a better balance between the different seasonalities of the different crops, and less dependency on the price of any one single crop.[36] For some, these advantages proved important during the 'rubber crisis'. The availability of lumber trees which can be cut and sold whenever needed (e.g., to meet a wedding expenses or in case of emergencies) is important. The advantages these holders mentioned reveal them to be risk-reducers. These advantages are of key importance because the holders' households are highly vulnerable and have difficult access to markets due to their remoteness (see above). However, there are also disadvantages attached to mixed cropping. One is the shade in plots containing rubber and other crops. According to experts from the Rubber Board, rubber trees are further affected since the availability of nutrients is diminished. However, most holders mentioned the difficult trade-off to be made between managing mixed crops properly and having household members working off-farm. This means that some homesteads seem to be managed suboptimally, affecting the yield of the different crops and this disadvantage is directly linked to the household members having diverse occupations. On the other hand, there are advantages in the strategy of agricultural diversification: in the case of livestock-rearing households, the dung can be used for manure.

The fact that household members also work as tappers and part-time agricultural labourers has both a positive and a negative impact on these holdings. The positive aspects are the regular income they bring in from these jobs and the fact that these jobs can be easily accessed in times of need, especially skilled jobs (this was a strategy for some during the 'rubber crisis', see section 6.6.1). Another advantage is the potential for an interdependent relation to be forged between the employer and the tapper. In the long term, this relationship can lead an employer to help his tapper when the latter is in a difficult or emergency situation.

Alongside these positive aspects, employment as a tapper can have potentially negative sides. Irregular tapping due to uncertain weather conditions affects the work on the own plantation and homestead (Box 6.1). Furthermore, tappers also often mentioned the physical problems (due to a heavy workload), stressful nature of the tapping job and the boredom of always doing the same work.

The average income of holdings of Type II is twice that of Type I, even though both have similar areas of land and numbers of trees (Table 6.4). This is certainly the most important impact of diversified incomes from non-agricultural sources. Thus, these holdings do not suffer the pressure of having to derive a 'coping' income exclusively from agriculture. The fact that they are not dependent on an income which varies with the uncertainties and seasonalities of the agricultural sector is an advantage. However, for some holders, the combination of both agricultural activities at home (i.e. especially the morning tapping) and off-farm wage-employment proves to be a burden and leads to suboptimal management of the rubber. Some of these farmers mentioned that if possible they would like to invest in more land under rubber. This is interesting and is directly related to their considerable non-agricultural income; in fact, the money they are able to save can be invested in rubber from which they gain further income without, however, having to maximize the income from it; it acts as insurance against emergencies.

Though most holdings of Type III specialize in rubber, some pursue other income activities, including starting new businesses. This portfolio of activities has the following consequences: first of all, dependence on income from rubber is (somewhat) reduced with the new businesses bringing a little more independence from the fluctuations of the natural rubber price. Still, these holdings' income is largely dependent on the movements of the rubber price; secondly, owning land with rubber trees reduces the vulnerability of these holdings and improves the households' access to loans from banks, since land documents can be deposited. When faced with emergencies or big expenses (e.g., for a wedding), these holders clear-cut rubber plantations and sell the wood or, if this is not sufficient, a part of their land as well. Here, the salvage value[37] of the rubber trees is of great importance. Owning land improves one's access to bank loans since land documents can be deposited as collateral (section 7.2.4).

The impact of diversification on Type IV holdings is linked to the reasons that drove these farmers to get jobs as civil servants, i.e. the wish to get a good, secure job. A second impact of a steady civil service job or of self-employment is the reduction in the risks of agricultural production. This is because the income from rubber (or from mixed crops) does not need to be maximized. There are, however, negative impacts as well: These holders have to employ tappers to harvest their latex, but have very little time to assist them and control their work, so they often face difficulties with their employees. Mistrust and dissatisfaction were often cited as negative aspects, as well as suboptimal management of rubber farms.

For Type V holdings, the impact of their diversification activities is generally positive; these households have a sufficient income to make a comfortable living and can also withstand shocks. There are, however, two important constraints: for some farmers who are leasing land to diversify into cash crop production it is difficult to find skilled tappers and they predict that this situation will only get worse in the future. The multiple occupations they have, not only on their own farms but also on the leased land, at times becomes quite a strain and even, for some, a reason to stop leasing land. One common outcome of the diversified and diversification strategy of the 'bigger and richer' rubber holdings (Types IV and V) is that members of the younger generation growing up in these households are normally not interested in following agricultural activities in the long term. The high-level education of these people enables them to pursue further studies and find well-paid jobs in the (non-agricultural) private or government sector. This does have an impact on the long-term planning of the farm activity as the family will need to decide at some point whether to increase or reduce rubber activities depending on whether at least one child is interested in taking over the farm's activities or not.

## 6.8 SUMMARY

This section first summarizes the status of the holdings in terms of their diverse activities and their progress towards diversification (section 6.8.1). Then, it summarizes the major driving forces that have resulted in this diversified status (section 6.8.2). Finally some thoughts on the social mobility and potential pathways of these holdings (section 6.8.3) are presented.

### 6.8.1 Status of Diversification within Rubber Holdings

Analysis of the income activities of different types of rubber holdings in Thalanadu shows that they base their livelihoods on a heterogeneous portfolio of activities. The most diverse portfolios are found at the extremes of the size range of holdings, i.e. in the most marginal holdings and in the biggest holdings analysed. These are also the poorest and the richest holdings in Thalanadu. Only the small holdings with a medium income, in the middle of the size range, are less diverse and specialize more in rubber cultivation.

Income sources are diverse both within and between the agricultural and the non-agricultural sectors. The agricultural sector plays an important role in the income of all types of holdings, but this depends to a great extent on the size of their holding and rubber plots. Natural rubber provides a direct income for all of them and an indirect income for the smallest ones, since they also work as tappers and labourers on other plantations. However, the income from other cash crops is a further important source of cash, especially for the richest ones who have planted cash crops on leased land. For the poorest holdings, mixed crops including staple food crops allow a certain level of subsistence. Income from jobs in the manufacturing and services sectors (i.e. the non-agricultural sector) is not important across all holdings, but it does at certain times play a key role for certain holdings. A key determinant is gaining access to such jobs in one of these sectors, as well as who the employer is and the terms of employment; this varies greatly across holdings. Households with a member in a long-term government post (Type IV, some in Type II) are far better placed than those with members in part-time employment in the private sector (most in Type II). For households with their own businesses (self-employed), the income situation depends on the type of business. Normally, marginal holdings can start only small and part-time activities (e.g., petty trade) while larger ones can take the risk of venturing into more important activities (e.g., building a smokehouse to grade and trade rubber sheets).

The process of diversification has been analysed for all holdings between 1995-6 and 2003-4. This includes the period of very low natural rubber market prices, commonly known in the field as the 'rubber crisis'. Interestingly, diversification as a process was less marked than expected. This means that holdings were already diverse before the financial year 1995-6 and that only a few farms further diversified after that. The 'rubber crisis' itself did not provide a reason to embark on new activities. In fact, to survive this period, reducing expenses proved to be more important for coping than diversification. Diversification towards the creation of opportunities in the manufacturing and service sectors took place only in selected bigger holdings, on which entrepreneurial farmers decided to invest in new activities. The aim of this study was not to analyse the process of diversification in the long term. However, farmers' accounts of the history of the cropping pattern (section 5.1.1) suggest that the cropping pattern has mainly changed towards more cash crops and rubber plantations, while the cultivation of staple food crops has diminished. Non-agricultural opportunities appear to have become progressively more important over recent decades.

### 6.8.2 Major Driving Forces Towards a Diversified Income

Each type of holding with a diverse source of income has a different set of driving forces which have resulted in its status. Analysis suggests that for the marginal holdings (Types I and II), vulnerability is the key driving force. Vulnerability is firstly caused by the low monetary income of these holdings due essentially to their small area of land. However, their vulnerability does not depend only on the size of their income, but also on the seasonalities of crops, climatic uncertainties and other external threats such as price fluctuations. Thus, these holdings are forced to generate their income from different sources in order to balance all these factors as far as possible. They do minimize their risks. Also, the cultivation of crops for subsistence has to be seen in this way. Since most income is spent as soon as it is earned, regularity of monetary income is a key feature. Marginal holdings get into precarious situations, especially when income drops, for example, because of the lack of agricultural employment during certain seasons. Many households are then forced to get by on loans from formal and informal sources (section 7.2.4). The fact that they diversify mostly within agriculture does not reduce their vulnerable position, since their dependence on the uncertainties of agricultural production remains. Thus, vulnerability can only be reduced through job opportunities outside the agricultural sector, and this has happened in some cases (especially in holdings of Type II).

Driving forces on bigger holdings are different than on marginal ones. In fact, large holdings (Types IV and V) have a diverse income portfolio because they have opportunities of access to these diverse activities. Leasing land, which many of the largest holdings do, is one example: leasing land requires a high availability of capital and the capacity to resist in case the venture fails. Investment in other activities in the manufacturing and services sectors also depends on the household's capacity to finance these activities, either from their own savings or with loans from the bank. They do, however, have motives other than simply a desire for greater income. Indeed, such business initiatives are often linked to the desire to reach or preserve a certain social status.

Finally, the important positive income characteristics of natural rubber as a crop should not be forgotten when analysing the driving forces. Although income from own rubber does not form the main income source of the most marginal holdings, rubber does provide a large share of the direct income along with the indirect income from work on other farmers' plantations. Marginal farmers can diversify their income only because bigger plantations exist nearby.

### 6.8.3 Diversification and Social Mobility

The present work focuses on the analysis of the income situation of the holdings at a particular point in time. Following this, holdings have been classified into different types. However, it is important to note that this is not a stable situation but can change along with income activities, both expected and unexpected, planned or unplanned. Also, some types of holdings demonstrate a similar income pattern (e.g., focusing on natural rubber production and other agricultural incomes, see Types III and V), but they have very different asset situations as their points of departure. While some of them have enough land to cover all their household expenses (Type V), others have to minimize their labour costs in order to make a living (Type III). Thus, the 'point of departure' of the holdings is important.

The analysis of the upward and downward social mobility of the different households shows the following: the marginal holdings (Type I) have, as expected, many constraints which make upward mobility rare. Their constraints are that they have almost no capacity to save in order to invest in new land, new crops or other businesses, nor to access loans. Instead, they are faced with the risk of downward mobility because of their lack of insurance and savings which can guarantee a regular income in times of crisis or emergency. Although informal short-term 'safety nets' exist, what the long-term consequences might be was not in the ambit of this study. However, on some poorer holdings which had started with almost no assets, an interesting strategy could be observed. Through constantly buying and reselling land and houses over many years, some households had managed to accumulate at least enough land to be able to plant a rubber plantation and build a house.

On medium income holdings (Types II and III), it was observed that many households would like to buy more rubber plantations in the future. Due to the good income rubber provides and its low management requirements, it is as important a crop for them as it is for holdings focusing on non-agricultural income (e.g., business). While the former can substantially improve their income through new plantations, the latter can, with minimal inputs, get at least a basic income which covers a part of the household costs. However, the latter's main income still comes from their paid employment. For both, the salvage value of the plantation and the value of the land as a capital goods are important in cases of emergencies or life-cycle events. These medium-income farmers are judged to have a stable livelihood situation; however, in the future, they will probably still base their livelihoods on the same activities they have at present.

Finally, some farmers of Types IV and V showed the potential for upward mobility. Some holders who have managed to start new business activities in the manufacturing and business sectors said that they would soon have a larger income from these activities than from rubber. They would thus move from being mere farmers to becoming 'business families', a higher social status. This fact was important for certain holders who did not want their children to be dependent on the uncertainties of agriculture. They also said that their well-educated daughters and sons were not interested in pursuing agricultural activities in the long term. Rather, these young people would prefer to look for jobs in the private or public sectors, which often results in their having to migrate to the cities or to other Indian states, and sometimes even abroad.

## NOTES

1. This section reflects the views of the interviewed rubber holders, and differs thus from the section on the general cultivation practices of *Hevea brasiliensis*.
2. On the consequences of pineapple plantation as an intercrop in rubber, see Véron (1998: 137-9).
3. Tapping panel dryness, also called brown blast, is a disease which leads to the partial drying up of the tapping cut. It is considered a physiological disorder of the tree, and is not caused by any pathogen (Rubber Board 2003b: 54). According to interviewees, panel dryness can be caused by too frequent tapping or by tapping during the rainy season.
4. Rainguards are protective panels fixed above the tapping cut to prevent water from flowing into the latex vessel during the rainy season.
5. The following is a concrete example of a leasing contract of a farmer who leased out his rubber land for slaughter-tapping: the farmer leases his land with all the old trees for two and a half years. During this time, the entire

latex yield belongs to the person leasing the land. The leasing rate has to be paid in three instalments, one at the beginning of the agreement, one after the first year, and the last one at the end before the cutting of the trees.

6. In Kerala, the financial year runs from the beginning of April to the end of March of the following year. Since this corresponds very closely to the rubber season, the financial year is taken as the reference year.
7. See Table 2.1, section 2.2.3. Of importance for this case study is the fact that the income from rubber sheets is divided into (1) an agricultural income from latex, and (2) a non-agricultural income from the processing of latex into rubber sheets. In fact, strictly speaking, processing of agricultural products cannot be taken as agricultural income, but as a service.
8. To be exact, the categories are: < 1 ha, 1-1.9 ha, 2-3.9 ha, 4-9.9 ha. Since the study location contains no holdings larger than 10 ha, bigger categories do not apply.
9. Normally, a tapper will tap on alternate days on his own farm and on his employer's farm. However, due to circumstances such as rain, this rhythm can be postponed, and it can thus occur that he cannot tap his own trees for four or five days, and thereafter on two consecutive days. According to some farmers, this irregular tapping is not healthy for rubber trees.
10. For example, in one household a woman said that she did not want to plant rubber in one of the household's plots because her father planned to hand that plot over to her sister; and since her sister lives in town, she would probably sell it to somebody else. Hence, planting natural rubber on that plot would be unwise.
11. In Thalanadu, most tappers are not organized in trade unions. However, if the price for natural rubber stays at a good level for some time, tappers will start bargaining for a rise in salary, Rs. 2-5 per 100 tapped trees. Since most tappers will ask the same at a certain period, employers have to accept the demand. Another way of bargaining is as follows: tappers, who, due to the limited size of the plantation, cannot tap the 300-50 trees that represent the standard quantity in Kerala, will say that their effort to tap 150 corresponds to that of somebody tapping 200 trees. Therefore, they will ask for a salary corresponding to that amount.
12. All the names in the boxes have been changed for anonymity.
13. It was noted that for some farmers work as *kooli* is not an option as their family status does not allow them to be employed in any agricultural work other than tapping. This fact is further detailed in the institutional analysis in section 7.2.6.
14. Bringing the latex from the plantations to the farmhouses was stated by many as being very tough work, especially in plantations located in hilly and steep areas. This was stated by many tappers as being the reason for physical pains such as backaches.

15. During the fieldwork for this study, there was ongoing construction work on a new road in Choovoor. This road was built under a central government food-for-work programme which promotes the employment of local people who are paid in both cash and kind.
16. Many farmers mentioned, for example, that streams which previously had water the whole year are now often dry for a couple of months each year.
17. For the same agricultural jobs, for example clearing land, a woman earns around Rs. 65 per working day while a man gets around Rs. 120.
18. In a very few cases women also mentioned that they sometimes tap the rubber trees. Though this happens, it still remains rare in the location under study.
19. Costs refer to the 'costs' of transporting goods on foot, sometimes to reach bigger centres by public transportation.
20. One woman said that, to reach her part-time work in a tea factory, she has first to walk 45 minutes, and then take a 20-minute bus ride. In one day she thus spends more than 2 hours travelling.
21. According to certain interviewees, water availability has changed drastically over the last years. There was a heavy drought during the field research in spring 2004; lorries had to bring water from far away to provide the local population with safe drinking water.
22. Regular income from a non-agricultural source also proved to be an advantage when applying for loans to buy land.
23. It was seen that some of these farmers did test a combination of natural rubber with other crops such as coffee, cardamom or pepper.
24. Due to the very high price of vanilla in 2003-4, vanilla seedlings were also very expensive.
25. According to the field survey, temporary government employees have a very low salary. A government sweeper working only in the afternoons gets around Rs. 3,000 per month, while a BSNL electrical technician working almost full-time gets only around Rs. 1,500 per month.
26. Some examples of civil service jobs are officer in the Indian Border Security Force, auditor general in the central government (but based in Kerala), head of a survey department branch, sweeper at a branch office of the Life Insurance Company of India.
27. The income management of these farms proves the same: income from rubber is often seen as an income used for 'extra expenses' (for example, for gifts or contributing to the wedding expenses of poorer neighbours), while the income for the household comes from paid employment.
28. Leasing of land for cultivation is prohibited under the Kerala Land Reform (Amendment) Act of 1970. However, micro-level studies suggest it is a common practice (Nair and Menon 2006).

29. At the time of the interviews, one farmer was planning to invest in a broiler chicken business. Analytically, this agricultural business would fit in the same category as leasing activities since it also requires important investments (stable, young chickens, special fodder) which can only be made by a richer farmer.
30. This problem relates to the encouraging fact per se that the children of agricultural labourers have a much better education and thus higher expectations from their jobs.
31. Some employers mentioned that since it is difficult to find tappers, it is worth keeping skilled tappers employed even during periods of low prices.
32. However, in 2003-4, their employment was reduced to a minimum; in fact, because of the high rubber price, fewer farmers would clear-cut their rubber plantations.
33. In India, this is a meal which includes meat.
34. 'Gold loans' are loans that one can get in exchange for depositing one's own gold in the bank.
35. An interviewee mentioned that at that time shops accepted debts up to around Rs. 4,000-5,000.
36. Some farmers stated that there are environmental advantages to planting rubber (e.g., better long-term soil fertility) but this could not be verified.
37. The salvage value is the residual value of a plantation once tapping has ceased. Normally, this equals the value of the wood from the rubber trees.

CHAPTER 7

# Role of the Local Institutional and Organizational Setting

The first section of this chapter describes the institutions and organizations that are relevant for rubber holders, followed by an analysis of the role of selected institutions and organizations by type of holding. The final section summarizes how they support or hinder the development of the income and diversification strategies of the different types of rubber holdings.

## 7.1 LOCAL INSTITUTIONAL AND ORGANIZATIONAL SETTING OF RUBBER HOLDINGS

While the analysis in chapter 6 was restricted to holdings in Choovoor and Maravikkallu, this chapter deals with institutions and organizations present not only in these two wards but also at what is called the *local* level, henceforth referred to as the *local institutional and organizational setting*. The necessity to broaden the geographical area from the wards to the *local* level stems from the fact that it is not possible to restrict the action and impact of institutions and organizations to a small area like a ward. Following Uphoff (1993: 609), *local* refers approximately to the area served by a rural market town. It includes a set of communities which have trading, intermarriage and other cooperative links with one another, where people have some possibility of personal acquaintance and usually some experience of working together (ibid.). It is the area rural people normally refer to when asked where they come from. In this study, the *local* institutional and organizational setting refers to a geographical area which includes the wards of Choovoor and Maravikkallu. This area is comprised of institutions and organizations present in Thalanadu centre (section 5.1.3), but also in the main centres of the neighbouring Panchayats, Teekoy centre (in Teekoy Panchayat) and Moonilavu centre (in Moonilavu Panchayat). As such, the area is not confined to the borders of the Panchayats in question but encloses parts of each of them (Fig. 7.1).

When analysing the spatial mobility of rubber holders living in Choovoor and Maravikkallu wards, it becomes immediately apparent that one must consider the geographical area in the manner outlined above. In fact, rubber holders are oriented not only towards Thalanadu centre, but often towards other centres as well. This is because families living in Maravikkallu ward have easier access to Teekoy centre, a place which offers better commercial opportunities, especially for selling one's own agricultural produce.[1] However, people still have to rely on Thalanadu centre, since important offices—the Panchayat office, the agricultural office—and the Cooperative Bank are located there. Furthermore, some rubber holders from Maravikkallu still market part of their sheets in Thalanadu centre because of the strong relations they have with traders there (section 7.2.2).

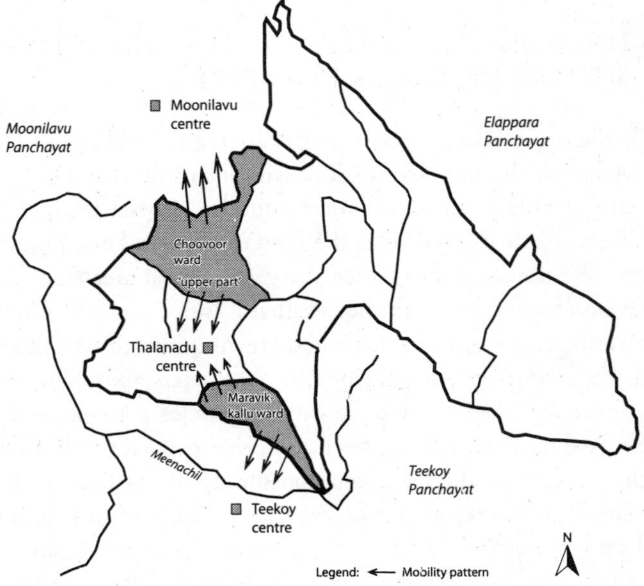


*Source*: Author's figure; not to scale.

Fig. 7.1: Map of Thalanadu Panchayat with main centres, and mobility pattern of farmers.

The situation is slightly different for people from Choovoor ward. Rubber holders living in the upper part of Choovoor travel to Thalanadu centre mainly for business and to deal with official matters at government offices. On the contrary, families from the lower part of Choovoor have more direct and easier access to Moonilavu centre (Fig. 7.1). Thus, they tend to buy and sell products there and go to Thalanadu centre less often, and then only for official administrative matters.

Limited spatial mobility and geographical factors do restrict rubber holders' access to the different types of institutions and organizations, as the following sections will demonstrate.

### 7.1.1 Local Institutional and Organizational Setting of Direct Relevance to the Income Generation and Livelihood Strategies of Rubber Holdings

This section describes the local institutions and organizations that interviewees mentioned as being of direct relevance to the income-generation and livelihood strategies of most rubber holdings. In Table 7.1, these are first divided into three categories: 'Organization', 'institution' and 'social relation'. The categories are defined in section 2.4.1.

The following sections will describe the functioning of and interactions between the different organizations, institutions and 'social relation'[2] in greater detail. First of all, there is a description and analysis of organization and institution related to rubber and agriculture; then, those involved in credit; and finally those with a social and welfare remit and institutions and organizations concerned with self-help and social networks.

*Rubber-related organizations and institutions*

The most important rubber-related local organization is the local field office of the Rubber Board, which is located in the centre of Teekoy.[3] It is the office of the local field officer, whose task is to implement the Rubber Board schemes the aim of which is to spread the plantation of natural rubber according to the scientific plantation methods promoted by the Board (section 4.4.1). The main promotional instrument is the subsidy scheme for replanting under the Rubber Plantation Development Scheme, which is implemented through the field officer. Besides this, the field officer organizes and gives training to farmers and tappers (on

TABLE 7.1: CLASSIFICATION OF LOCAL ORGANIZATIONS, INSTITUTIONS AND SOCIAL RELATIONS

| *Type* | *Organization* | *Institution* | *Social relation* |
|---|---|---|---|
| Rubber-related | Rubber Board field office; Rubber Producers' Society (RPS) | Rubber Board schemes; Rubber Board Price Stabilization Fund; Marketing of rubber sheets by RPS; RPS administered schemes; Rubber sheets market; Use of neighbours' rollers | |
| Others related to the agricultural sector | Agricultural office; Cooperative society; Thalanadu Panchayat; Farmers' Societies (e.g., for coconut, vanilla, *toddy*) Specialized traders (e.g., for spices) | Manure subsidy; Central Government-sponsored schemes; Panchayat-sponsored schemes; Markets for (non-rubber) agricultural products | |
| Credit related | Cooperative Banks (three branches); Federal bank; Agricultural development bank; Private moneylenders; Micro-finance schemes of self-help groups | Formal short-term credit market; Formal long-term credit market; Informal loans | |
| Social and welfare related | Panchayat office; Tribal office; Private school for SC/ST; | Housing loan scheme; Other schemes for below poverty line and SC/ST families; Reservation of government employments | Below poverty line families; SC/ST families |
| Self-help group and social network related | Women's self-help groups (SHGs); | *Kudumbasree*; Small income activities; Group loans from Panchayat | Women; Hindu Ezhava community; Social status |

*Source:* Author's interviews.

tapping, processing, and the use of rainguards), often in collaboration with the Rubber Producers' Societies present in the village. The field officer is also in charge of implementing more occasional and short-term schemes, for example for the development of tribal families[4] or the Price Stabilization Fund. The role and impact of the Rubber Board's replanting subsidy scheme implemented by the field officer on the rubber holdings is analysed the section 7.2.1.

In Thalanadu Panchayat there are four Rubber Producers' Societies (RPS): Thalanadu RPS, Maravikkallu RPS, Addukom RPS and Kalakodu RPS. Rubber Producers' Societies are promoted by the Rubber Board as a means of encouraging self-organization, especially of the poorest and most marginal rubber growers, who generally form the majority of their membership. According to the decisions of the Rubber Board, an RPS can be registered if it has at least seven rubber holders as members. Membership is open to smallholders who own at least 0.2 ha and a maximum of 5 ha of rubber plantation. Each member needs to pay an entrance fee of Rs. 50 and an annual fee of Rs. 10. One member of the Rubber Board—normally the local field officer—is a member of the committee of each RPS. However, he/she has no voting rights; his/her main task is to consult the RPS members. The field officer thus regularly collaborates with the RPS members (Mokhtar 2006: 59-60).

According to interviews with two RPS presidents in the wards, the activities that are implemented by the RPS depend to a great extent on the current political and economic climate, as well as on the energy and perseverance their presidents or committee members demonstrate in promoting activities. Between 1989 and 2001, the local RPS were active in distributing subsidized inputs (e.g., manure, pesticides) and materials their members needed for the plantations. These were subsidized by the Rubber Board between 1994 and 1998 with some of the funding coming from the World Bank Rubber Project.[5] Until 1998, subsidies were granted for the purchase of any inputs and materials, but from 1998 until 2001, subsidies have been limited to manure and spraying pumps. The material was sold at a 10-20 per cent discount at normal market prices, and the RPS got a 10 per cent commission. After 2001, the subsidy scheme was adapted and the RPS were excluded from its implementation. Since then, the RPS are mainly involved in extension but a few pursue some commercial activities (e.g., renting out pesticide spraying-machines owned by the RPS).

One RPS in Maravikkallu ward buys rubber sheets from its members once a week, on a specific day in order to remain active. The price is, on average, Rs. 0.25 per kg higher than in other shops (though informants said that some shops adapted their prices on that day). According to the president, some twenty to twenty-five of his members regularly sell their sheets through the RPS; this is not much, but enough to make running the activity viable at a small profit to the RPS. The sheets are then sold to a wholesale company, the Travancore Rubber Marketing Society, in Pala. This activity—besides generating some financial income—is important to the president of the RPS as it provides the RPS with a raison d'être. However, to buy the sheets they require a storeroom and this is an important constraint explaining why other RPS in the village are not involved in trading rubber sheets. But it is also linked to the fact that their presidents believe that farmers prefer to sell their sheets to rubber dealers because of the many advantages they get, such as pre-financing, different daily prices, acceptance of lower quality and longer opening hours. Presently, the RPS are mainly engaged in training and extension, as well as the promotion of smaller Rubber Board schemes such as the Housing and Sanitary Scheme for SC/ST rubber tappers.[6] Some RPS also give an award for the best students amongst their members' families. Interviews with farmers showed that many could not quote any advantages of membership of an RPS and had therefore not paid their membership fees for many years. Furthermore, some of the most marginal farmers were not even aware that the RPS existed. As the Rubber Board believes in the future potential of the RPS to process and market natural rubber, an awareness-raising campaign was dedicated to the strengthening of the RPS in 2003.[7] During the campaign, Rubber Board staff members in most of the RPS in the country organized seminars on the topic. However, while one of the RPS presidents interviewed was well aware of the potential of collective processing and marketing, another one did not believe that this alone would provide sufficient motivation for farmers to work together.

The marketing of rubber sheets is another important *institution* in the natural rubber-producing sector. Most farmers sell their rubber sheets to local village rubber dealers, all of them licenced by the Rubber Board. Some of the larger farmers are an exception as they stock their rubber sheets and then take bigger quantities of them to wholesalers in the nearby town. Due to institutionalized marketing practices between rubber dealers in the village and the rubber farmers, the process of marketing covers more than simply 'selling sheets'. The consequences of this for rubber holders' income generation and livelihoods is discussed below (section 7.2.2).

Besides the three main rubber-related organizations and institutions described earlier (the Rubber Board field officer, the RPS and the marketing of rubber sheets), two more need to be mentioned. The first of these is the Cooperative Bank in Thalanadu, which is currently the authorized distributor of subsidized inputs (manure, fungicides) to rubber farmers who receive (re)planting subsidies from the Rubber Board. The second is the institutionalized practice between neighbouring rubber farmers of sharing the rollers used for preparing the rubber sheets. Many of the bigger rubber farmers who have their own rollers allow their neighbours (normally marginal holders) to use them. In exchange, they are expected to give each year one or two days' yield of rubber sheets to the owner of the roller. According to the interviewees, this practice functions very well in the Panchayat. Furthermore, in Choovoor, three to four community rollers have been installed by the Panchayat for SC/ST families. These can be used free of cost but have to be maintained by the community.

*Other agricultural sector-related organizations and institutions*

The agricultural officer, is responsible for managing the agricultural office located in the Thalanadu centre. Thus, he is the key person—along with his office assistants—in the implementation of the many governmental schemes in the agricultural sector (e.g., plantation subsidies for specific crops, land terracing, irrigation or temporary schemes such as the Drought Release Fund implemented in 2004). The agricultural officer works under the Department of Agriculture; in each Block Panchayat (a political-administrative subdivision of the district) there are eight to nine agricultural officers, each of whom employs agricultural assistants and office assistants. Agricultural officers are normally responsible for only one Panchayat but in certain cases they may be in charge of several. They implement different types of schemes: schemes sponsored by the central government (known as *Centrally Sponsored Schemes*) for the development of specific crops (e.g., rice, tuber crops, coconut, and spices)[8] and State Sponsored Schemes, some of which are developed in collaboration with the Panchayat. Table 7.2 gives examples of some of these schemes. In 2003-4, Thalanadu Panchayat sponsored four schemes. These schemes were developed in collaboration with, and implemented by, the local agricultural office. They were the promotion of vanilla cultivation, subsidization of irrigation pump sets, subsidization of digging open wells, and the promotion of automated rubber-drying units.[9] Each scheme requires farmers to contribute to the costs. In the financial year 2003-4, around Rs. 1,200,000 were spent on these schemes, of which Rs. 800,000 came from the farmers themselves and Rs. 400,000 from the Panchayat. Section 7.2.3 will look more closely at the functioning and role of one of these schemes, the Vanilla Promotion Scheme.

Besides implementing schemes, the agricultural office organizes regular training sessions and seminars, sometimes in collaboration with the Commodity Boards (e.g., Coconut Board, Spices Board) and provides information on how to cultivate different crops. However, its main work is in implementation. Thus, at specific periods, up to fifty farmers a day come to the office to fill in forms or collect subsidies. According to the agricultural assistant, all farmers are informed about the schemes and

TABLE 7.2: SELECTED AGRICULTURAL SCHEMES IMPLEMENTED BY THE AGRICULTURAL OFFICER IN THALANADU PANCHAYAT, 2003-4

| *Sponsor* | *Name of scheme* | *Type of scheme* |
|---|---|---|
| Central Government | Pepper Fund (from Spices Board) | Free pepper planting material; subsidized manure |
| | Coconut Development Fund | Subsidy per plant; minimum of 15 trees to be planted |
| State Government | Drought Relief Fund | Compensation-related to losses due to drought |
| | Vegetable Scheme | 50 per cent subsidy for seeds and manure |
| | Banana Cultivation Fund | Free manure for a cultivation of a minimum of 0.5 hectare |
| | Terracing/Land-Scaping Fund | Subsidy of labour costs of terracing |
| Thalanadu Panchayat | Vanilla Promotion Scheme | Subsidy for seedlings: Rs. 1,500 per 100 metres of seedlings |
| | Irrigation Pump Subsidy | 50 per cent subsidy for pumps (Rs. 2,500 per applicant) |

*Source:* Author's interviews.

have good access to them. She also mentioned that agricultural assistants help farmers to fill in forms if and when required. No special agricultural schemes have been developed for SC/ST families. In conclusion, it is important to mention that the agricultural officer does not normally deal with questions related to rubber cultivation; in fact, it is agreed between the agricultural officer and the Rubber Board field officer that only the latter will give advice to rubber holders about rubber. Both officers confirmed this. In this sense, the Panchayat's promotion of rubber-drying units (see above) is an exception.

The Farmers' Societies are another important agricultural-related organization promoted by the Panchayat and the different Commodity Boards. In the two wards, many farmers mentioned that they were members of Farmers' Societies for specific crops, for example, the Coconut Farmers' Society, the Milk Society, the Vanilla Farmers' Society (around 30 members), or the Toddy Farmers' Society (around 150 members, including from surrounding villages). These groups are active in the promotion of the specific crops, organizing information exchange, field visits and other activities, but they also lobby for better marketing conditions (e.g., the local price of toddy was said to have gone up quite considerably thanks to the actions of the Toddy Farmers' Society). However, some of these groups are also instrumental in obtaining favourable conditions for group loans from the local Cooperative Bank (section 7.2.4).

People involved in the marketing of local non-rubber agricultural products (traders, local shop owners) also play an important role in the livelihood strategies of rubber holders, especially those cultivating mixed crops. Traders normally specialize in trading a single or a small number of commodities. One trader dealt only in cloves (*gramboo*) and nutmeg (*djadi*), while another bought only coffee and pepper. The following details were provided by a trader specializing in cloves and nutmeg.[10] In and around Thalanadu centre, where the trader is based, there are around ten traders dealing in cloves and nutmeg. They live in the centre and, besides trading, are also farmers and rubber holders. They operate as traders only in the harvest season, which is once or twice a year depending on the crop. They usually buy the products directly from the farmers and their prices depend on the quality available. Some farmers do, however, enquire (by phone) about the price and then bring the product to the trader's house themselves. Before selling the spices to bigger traders in Eratuppetta (the nearby town), the trader will grade the product according to quality. The trader also said that many farmers have planted new nutmeg trees in the Thalanadu area during the last five years. This is due both to the current good prices for nutmeg and to the fact that it can be harvested twice a year.[11] According to the trader, nutmeg is the villagers' second-best source of income after natural rubber. The high price of nutmeg in recent years has forced larger traders to come from Eratuppetta to buy nutmeg directly from the farmers. These bigger traders normally offer a better price than the smaller ones are able to. Local traders can only resist this competition by maintaining a personal relationship with the farmers and hence often buy from the same farmers and pay them regular visits.

*Credit-related institutions and organizations*

Almost all those interviewed mentioned those organizations that provide savings and credit facilities to rubber holders. Formal organizations such as the Cooperative Banks and the Federal Bank in Teekoy were often mentioned. Many rubber planters deal most regularly with the Cooperative Bank (which has branches in Thalanadu, Moonilavu and Teekoy). This is because many government (and local Panchayat) agricultural schemes are administered by the Cooperative Bank. For example, the Cooperative Bank lends money to Farmers' Societies under a new scheme introduced by the Panchayat around 2003 (section 7.2.3). When this is considered together with the individual loan facilities, it is evident that the Cooperative Bank provides a wide set of services to the farmers. However, some rubber holders also mentioned other banks outside the local area, such as the Agricultural Development Bank (with a branch office in Eratuppetta), which they said gives loans under favourable conditions for long-term projects such as purchasing of agricultural land, starting a plantation (natural rubber, coconut, or others) or developing a new activity (e.g., milk production). But the bank does not have facilities for savings. It was mentioned that farmers use other formal institutions such as post office accounts for small and regular savings.

Apart from these formal institutions, interviewees did mention more informal ones. There are private moneylenders who occasionally work with farmers.[12] Farmers also often make use of small micro-finance schemes promoted by self-help groups (see below) to finance small and unexpected expenses. These funds make it possible for many marginal holders to bridge seasons when employment opportunities are scarce or inexistent (section 7.2.4).

*Social and welfare organizations and institutions*

There are various social and welfare organization and institutions available to different people and at different levels. The most important scheme is the Housing Scheme of the Panchayat which gives people subsidies to build their own house. This scheme is reserved for people on very marginal income, i.e. those living *below the poverty line* (commonly abbreviated as BPL).[13] To qualify for the subsidies, families have to apply to the Panchayat office via the *ward member* (the elected person representing the ward). A working group of the Panchayat, the *Karma Somadi*, then evaluates all the applications, visits the applicants and makes its recommendations. Based on this, the *Grama Sabha* (the assembly of all the inhabitants of the ward) discusses the list and decides on the allocation of subsidies. The final decision is then taken by the *Panchayat Committee* and it is customary that all the families recommended receive a subsidy. If there are insufficient funds available one year, some families will receive a subsidy at the next opportunity. In Thalanadu Panchayat, around ten BPL families get Housing Scheme subsidies each year. The construction subsidy is generally Rs. 25,000, paid in three instalments: the first when the foundation of the house is finished, the second when the walls are built, and the final one when the roof is finalized. Subsidies for maintaining a house are smaller and on a case-by-case basis. Many of the most marginal holders interviewed said that they had received subsidies to construct their house, a fact which was apparent. Though it is not the purpose of this work to analyse this scheme, some advantages and constraints are nevertheless worth mentioning. The fact that many people live in houses subsidized through this scheme leads to a first positive conclusion on its impact: without the assistance of the Panchayat, many of these families would probably still be living in low-quality huts. However, many households complained about how the scheme is implemented: the high costs linked to the complicated application process (e.g., proving that the land belongs to a family member and also paying bribes to certain officers) are a major problem. People who wished to construct a house in remote and hilly areas complained that the subsidy did not take into account the high costs of terracing and transporting material to the construction site. In sum, the impact of the Housing Scheme varies from household to household but its general developmental impact is positive.

Besides the housing scheme, the Panchayat also subsidizes solar lamps and solar energy panels for households without electricity. This was the case in Choovoor ward. Subsidies are also available to build latrines.

Special schemes exist for Scheduled Castes and Scheduled Tribals (SC/ST). Some of the houses of SC/ST families were constructed with the help of the Housing Scheme of the Tribal Department, financed in equal parts by the local Tribal Department (50 per cent) and the Block Panchayat (50 per cent). In other SC/ST households, people can take special training (e.g., in stitching) and subsidies to buy tools (e.g., for sewing machines). Children from SC/ST families can also attend subsidized boarding schools, which proves helpful if parents have to migrate regularly in search of work. SC/ST households have to show a caste certificate to get access to these schemes. The most important of all these schemes for people from SC/ST households is the reservation of a certain percentage of government jobs for them, and yet, statistically, the chance of getting such as job is quite small.[14] Nevertheless, some families in the ward under study did find employment in the civil service and with it all the advantages already mentioned. Interestingly, a local expert mentioned that a comparatively high percentage of SC/ST families in Thalanadu have a member working in the public sector. This appears to be linked to there being a special private school for people from SC/ST families (section 7.2.5). The fact that some young people from rubber households aim, come what may, for a civil service job also has its negative aspects: respondents from different families said that some young people do not look at other ways of earning money for the family as they expect to get a job in the civil service soon. However, in the meantime, all the other family members have to work hard to bring sufficient income into the household.

Looking at the social and welfare institutions and organizations, one can also recall the Rubber Board schemes for poorer tapper families (e.g. the Housing and Sanitary Scheme for SC/ST rubber tappers), implemented via the RPS.

*Self-help groups and other networks*

In Thalanadu, many women are members of women's self-help groups (SHGs). These SHGs follow different aims, the main one being to further exchanges between peers, to empower women through the promotion of small income-generating activities (often group businesses) and to run different savings and credit schemes (e.g., *chitties*[15]). Furthermore, SHGs are good platforms to disseminate relevant and useful information, for example, about available schemes. Many of these SHGs are nowadays affiliated to *Kudumbasree*, a state project implemented through the Panchayat office; once these groups have more than ten registered members, the Panchayat supports them with a group loan (normally Rs. 10,000) to be distributed amongst the members or invested in common activities (e.g., one group started planting bananas). However, the different micro-finance schemes have shown themselves to be the most attractive activities. In fact, during interviews, rubber holders (who were often the husbands!) praised the benefits of the small but unbureaucratic loans given regularly to their families during seasons when there is no other income. One SHG managed to open a small consumer goods shop in Choovoor; women from the group were regularly employed part-time in the shop and derived a small income from it.

While participation in most SHGs is independent of caste or religion, other groups have these as their basis. In Thalanadu, for example, a group of people belonging to the Hindu Ezhava community meets regularly for specific activities such as stitching. In the Christian community, members within the same neighbourhood have a monthly prayer meeting. Coffee and tea are served afterwards along with some sweets as part of an informal get-together.

In some households, church and religious practices also have advantages for off-farm businesses. For example, one person living on a rubber holding uses the daily mass in the local church to get orders from her neighbours. As she does the stitching work at home, located in a remote place, after the mass proved to be the best time and place to meet people, discuss with them and take orders. She then brings the finished dresses to the same place some time later. Interestingly, many orders are linked to festivities like Christmas, Easter and Onam,[16] as people order many new clothes for these occasions.

### 7.1.2 Institutional Support During the 'Rubber Crisis'

Was there specific organizational and institutional support of the rubber holdings in the Panchayat during the 'rubber crisis' between 1997 and 2002? This study has not identified any new schemes introduced specifically because of the crisis. However, rubber holders used existing schemes differently at different moments of the crisis. Around 1997-8, many farmers asked for subsidies under the Rubber Plantation Development Scheme, since they had decided in the previous year to renew their old plantations and take advantage of the very high rubber wood prices. However, at the peak of the crisis (in 2000-1), there was a reduction in the number of applicants, and only after 2003 did demand for financial support start to rise again.[17] Another measure taken by the Rubber Board was to change the amount of the subsidies given to the farmers (section 7.2.1). During the 'rubber crisis', the central government decided on the most important measures, introducing trade control and price support mechanisms (section 4.4.2). Only one measure was developed during the crisis, namely, the Price Stabilization Fund, which came into force on 1 April 2003. The main aim of this fund is to insure rubber holders (with less than 4 ha) against a sudden drop in price.[18] However, according to the field officer, only thirty farmers of the Panchayat were participating in this fund one year after the introduction of the scheme. He explained the low participation by the fact that rubber prices were rising. However, during the interviews with rubber holders, it was seen that only a few of them were aware of this fund.

## 7.2 ANALYSIS OF SELECTED LOCAL INSTITUTIONS AND ORGANIZATIONS

This section describes and analyses the role that some institutions and organizations play in promoting or hindering the diversification of income-generating activities. It looks at those institutions and organizations that have been regularly mentioned during interviews as having positive and negative impacts on the livelihoods of rubber holders. They have been selected based on the relevance of the institution/organization for income generation. They are analysed according to the different types of rubber holdings as described in section 6.2, since it can be shown that different institutions/organizations have different effects on the different types of holders. The institutions/organizations analysed are the Rubber Board replanting subsidies, the local rubber marketing institutions and organizations, the Panchayat Vanilla Promotion Scheme, loans from informal and formal organizations, social networks and the social status of rubber holders.

### 7.2.1 Rubber Board Replanting Subsidies

As mentioned above, the Rubber Board is represented in the village by its extension officer, the field officer. The Rubber Board field officer offers support to rubber holders on all rubber-related questions. However, the most important scheme implemented by the field officer is the Rubber Plantation Development Scheme (RPD), which supports rubber planters with subsidies for establishing new rubber plantations and rubber replanting. The subsidy amounts to a total of Rs. 12,000 per hectare, and is disbursed during the first six years after replanting (the immature period) when the trees are still too small to yield latex. Planters also get additional subsidies if they use good quality planting material (commonly called poly-bag planting material). Besides disbursing subsidies, the field officer is also responsible for making an annual check of the young plantation; he checks whether weeding and manuring are being done and gives the permit to start tapping the trees (when the girth has reached a minimum of 50 cm). These subsidies are given only to growers with plantations smaller than 5 ha and subsidies are given for a maximum of 2 ha per year.

The payment of subsidies is based on strict criteria which aim to ensure that rubber is planted according to the Rubber Board's recommended practice. The criteria are as follows: firstly, the area to be (re)planted has to officially belong to a member of the family (legal certificates have to be produced); secondly, the planting distance has to be in line with the Rubber Board's recommendations, which are a standard of 420 (minimum required) to 500 (maximum allowed) rubber trees per hectare; thirdly, only approved planting material can be used; fourthly, rubber has to be cultivated as a monocrop. This final criterion means that only a minimal number of other non-rubber timber trees are allowed to be present inside a new plantation/replanting; all others have to be cut down. In practical terms, this means that a plantation of 420-500 trees can contain a maximum of 20 other (non-rubber) trees.[19] However, seasonal intercrops (e.g., pineapples) are allowed during the first years of replanting when the rubber trees are still small.

The analysis shows that this scheme has a variety of impacts on the different types of holdings. Most of the marginal holdings (Types I and II, with less than 1 ha) have no interest in these subsidies or have difficulty in accessing them or, in a few cases, have no information about them (Table 7.3).

The main reason many smaller farmers are not interested in applying for subsidies is that many of them have their rubber trees in plots with mixed cropping (mostly big timber trees); since these farmers make their livelihood from mixed crops, there is no advantage in removing any of these plants as is required by the Rubber Board. Above all, they do not want to cut down the big valuable trees in their plots. In fact, the crisis in the rubber sector at the end of the 1990s has shown that keeping a handful of timber trees standing can be an important source of immediate cash when no other income is available. Another problem is the density of the plantations of marginal holdings, which often does not follow the standards of the Rubber Board. It seems that in many cases, these farmers do not even get in touch with the field officer; many say they know that they will not get subsidies and thus feel that there is no need to enquire. A small number of marginal holdings had approached the field officer about subsidies, but they said that the time required and the bureaucracy (e.g., to prove that the land belongs formally to a family member) is too great a constraint. This is particularly the case for marginal holdings which are eligible only for limited subsidies due to the small size of their plots.[20]

Small and semi-medium holdings (Types III and IV, with 1 to 4 ha) do weigh the advantages and disadvantages of applying for replanting

TABLE 7.3: COLLABORATION WITH THE RUBBER BOARD FIELD OFFICE

| | *I* | *II* | *III* | *IV* | *V* |
|---|---|---|---|---|---|
| Know the Rubber Board (RB) field office | 85% | 60% | 100% | 80% | 100% |
| Already made contact with the RB field officer | 15% | 0% | 89% | n.a. | 100% |
| Received RB subsidies at least once | 0% | 0% | 78% | n.a. | 100% |

*Source:* Author's interviews.

subsidies, but most of them continue to apply as always (Table 7.3). During interviews, many rubber holders said that they had only applied because they would not have to cut down many other trees in their plots: either their replanting was almost free of other trees, or the field officer had allowed them to exclude from the calculation the part of the land on which there were big trees.[21] Some of them agreed to fell some trees because they were old and cutting them in the future would only damage the rubber plantation. But farmer who had many young timber trees said that they would not be applying for subsidies. Many interviewed farmers explained how they had reached this decision due to their experiences since the 1980s. During the 1980s—when the Rubber Board started the scheme and many farmers applied—the application of the rules and regulations was handled very strictly by Rubber Board staff and consequently most farmers had to cut down most of the timber trees on their land. During the rubber crisis of the mid-1990s, many felt that it had been a mistake to have felled the trees, as they could have generated crucial additional income. It is interesting to note that the Rubber Board seems to be more flexible in applying the rules today; for example, it accepts that certain areas with trees can be excluded from the subsidized area.

Type V rubber holdings (semi-medium and medium, with more than 2 ha) had experiences similar to their colleagues in Types III and IV holding. All of them applied for subsidies during the 1980s and have continued to do so, even though they are more critical towards them today. This is shown by the fact that many now refuse to cut down their timber trees even if they are formally requested to do so as a condition for receiving subsidies. Many calculate that keeping timber trees is more rewarding than getting subsidies; for example, one farmer explained that in the 1980s he received a total of Rs. 5,000 in subsidies for replanting 1.3 ha and had to cut down ten timber trees to be accepted in the scheme. He sold the only tree he had kept at the end of the 1990s for Rs. 23,000. This gives a broad idea of the financial loss incurred by farmers who had to cut down their trees.[22] However, it must also be said that most of these bigger holders work closely with the Rubber Board and its schemes. In fact, most of them plant rubber as a monocrop and get other benefits linked to these subsidies from the Rubber Board such as advice, training and subsidized inputs (e.g., for manure or seedlings).

In sum, the subsidies do support income levels during the years of replanting, particularly those of the bigger holders. After replanting, bigger holders can also expect good yields due to the development of a scientifically planted and high yielding rubber plantation. However, for the marginal and smaller units that are often engaged in mixed cropping, subsidies are less attractive and accessing them is sometimes difficult. Most farmers, independently of the size of their holdings, are critical of the rule that only a small number of non-rubber timber trees can be grown inside the plantation.

### 7.2.2 Local Rubber Marketing Institutions

Local rubber dealers and Rubber Producers' Societies (RPS) are rubber holders' main partners when it comes to marketing rubber sheets. As already mentioned, rubber trees have the great advantage of yielding during almost eight months of the year; the latex is then processed into rubber sheets on the farm and sold locally. Marginal and small holders (Types I and II, and some of Type III) market these sheets, on average, every couple of weeks, because they need the cash but also because they have nowhere to stock them themselves. By contrast, bigger holders (some of Type III, and Types IV and V) sell their rubber sheets, on average, only every two months during the tapping season. In this way, they try to take advantage of their greater bargaining power due to the larger quantity they offer. This works because they have a less pressing need for cash.

Rubber holders in the study location take their sheets to rubber dealers in Moonilavu, Thalanadu and Teekoy. There are about half a dozen active rubber dealers in these three locations and the difference between the price each pays is in the order of Rs. 1 to 2.5 per kg (of a total price ranging from Rs. 53 to 60), depending on their deals with the wholesaler and their distance to the main centres. Each farmer has his own strategy in choosing the dealer; some holders phone several dealers to check which one is offering the best price on that day. Other rubber holders prefer always to work with the same dealer, since this gives them some advantages (see below). Finally, a smaller number of farmers regularly take part of their sheets to their Rubber Producers' Society since they think that the RPS always gives a fair price (i.e. the one stated in the local daily newspaper)[23] and also because they want to sustain, at least to some extent, the RPS as a farmer-run institution. However, that certain RPS only take delivery of rubber sheets one day per week is a problem, which is the main reason why only a small quantity is sold through the RPS.

Interviews have shown that there are institutionalized practices between rubber dealers and rubber holders that are particularly crucial for marginal holdings. One practice is that rubber dealers regularly give cash advances to rubber holders. According to farmers, they can get up to Rs. 500 from a dealer at a time. Sometimes, they also get advances from different dealers. These advances are then deducted from the next sale of rubber sheets. Farmers said that these advances are important to bridge the gap during the low-yielding season, but also in case of emergencies. However, certain farmers have institutionalized this practice to the extent that they take regular advances every two weeks and then pay back everything every two months with their rubber sheets. While some farmers say that they can only get these advances if they always sell to the same rubber dealer; others say that they can get these advances from different dealers. However, interviews with the most marginal holdings showed that these mainly work with only one buyer. For this practice, it is important that the rubber dealer is not too far away from the farm. Thus, farmers in Choovoor prefer dealers in Moonilavu, while farmers in Maravikkallu tend to sell to dealers in either Thalanadu or Teekoy (Fig. 7.1). Another advantage of always working with the same dealer is that he will also agree, from time to time, to take sheets of a poorer quality for a reasonable price. A second institutionalized practice is that dealers will agree to take rubber sheets in advance from the holder, but the holder can then decide when he/she will be paid, at the price level of that day. This practice has advantages for both rubber holders and dealers. The advantages for the farmer are firstly that he/she can avoid the risk of having to stock sheets, which often leads to a deterioration in their quality (e.g., due to mould), and secondly, that the rubber holder can maximize his or her income by delivering sheets during the yield-intensive months and getting paid in the low-yield period (February to April), during which prices are normally higher. For the dealer, the advantage of this interdependent relationship based on mutual trust is that he can be assured of the regular delivery of rubber sheets from the involved holders.

To summarize, interviews show that the two institutionalized marketing practices between rubber dealers and rubber holders hold important advantages for marginal holdings especially (Types I and II), as they depend on a regular cash flow from the sale of their rubber sheets. The practices are therefore important in sustaining the income generation of these holders. For larger holdings these institutions are less important since their bigger plantations allow them to generate enough income for their daily household expenses.

### 7.2.3 Panchayat Vanilla Promotion Scheme

The possible impact of non-rubber-related agricultural institutions and organizations is shown by the example of the Vanilla Promotion Scheme.

Farmers in Kerala started to plant vanilla extensively shortly after 2000 as different bodies (such as NGOs, the Spices Board, but also the media) had promoted it as a crop with high yield and economic potential and well-adapted to the Kerala climate, especially along the Western Ghats. Between 2000 and 2003, the international and local prices for green vanilla beans were very high and the Thalanadu Panchayat office started to promote the cultivation of vanilla amongst the local farmer community. In 2002, the agricultural officer, who was responsible for implementing the Vanilla Promotion Scheme, invited local farmers to an introductory class in the cultivation of vanilla, including input from experts from the Spices Board. From 2002 onwards, the Panchayat subsidized the purchase of vanilla plant seedlings: if a farmer bought 100 metres of vanilla seedlings, the Panchayat reimbursed him Rs. 1,500. In each of the years 2002 and 2003, ten farmers received these subsidies.[24] According to the agricultural officer, around 10 ha of vanilla were planted in the Panchayat between 2002 and 2004.[25] Though this corresponds to only a very small percentage of the total agricultural area in Thalanadu (3.3 per cent), the intensively cultivated vanilla on a small piece of land can be an economical interesting cash crop for farmers. Since this scheme represents a potential income-diversification activity, it is interesting to analyse how different types of rubber holders responded to the Vanilla Promotion Scheme.

Interviews with rubber holders show that, in broad terms, there have been three types of reactions to the scheme: some farmers did not plant any vanilla (for different reasons), others started experimenting with a few vanilla plants to get to know the crop and a third group took full advantage of the promotion scheme and invested in vanilla as a new cash crop. Of the 40 farmers interviewed, 19 mentioned that they had planted at least a few vanilla plants. The majority of those who had not were on marginal holdings (Types I and II). In fact, many of them were confronted with one or more of the following constraints: firstly, the high price of vanilla in 2002 to 2004 also caused seedlings to become very expensive and for most the price was prohibitive. Secondly, marginal farmers lacked the necessary land to plant a new crop; they would have had to cut down other well-established crop trees to have enough land without too much shade suitable for vanilla cultivation. Thirdly, most farmers lacked the necessary water to plant vanilla successfully. Fourthly, the risk of price volatility was one more reason for skipping vanilla. The group of rubber holders who planted a few vanilla plants comprises farmers from all types of holdings (Types I to V). Their main argument for planting only a few vanilla plants is that they wished to test the crop. Some farmers who had already done this mentioned that they felt that they would need more water than was currently available for vanilla to thrive and they would therefore not be planting any more. By contrast, a third group of small to medium farmers (Types III, IV and V) used the opportunity of the subsidy scheme to embark upon a more systematic vanilla cultivation. Six interviewed farmers received Rs. 1,500 as a subsidy to buy seedlings and one farmer got a further Rs. 750. Five of them planted the minimum number of plants necessary to get the subsidies (i.e. 100 plants), while two of them decided to cultivate more vanilla (around 300 plants). Most of these farmers mentioned that they had to get more information before they decided to plant vanilla and that is why they visited neighbours' vanilla plots and well-known vanilla farmers in the area. They also got information from TV programmes and a CD-Rom developed by a local NGO. Some of them mentioned that they were expecting the agricultural officer to organize more seminars on the topic. These farmers noted that one important aspect of planting vanilla is how much suitable land is available, i.e. land without too much shade. However, after two years of cultivating vanilla, some said that lack of water had turned out to be an unexpected constraint. Thus, some farmers plan to use a water pump.

There are some practices worth mentioning when considering how farmers make use of subsidy schemes. One expert vanilla farmer mentioned that some of the farmers did not follow best practice when planting the seedlings: instead of planting seedlings of one metre length, they cut them into smaller pieces so as to buy less planting material for the same amount of subsidies. The result was that the vanilla plants did not develop as expected. But for some larger farmers, the income from vanilla was not the most important motive for planting vanilla. One farmer said that he plants vanilla 'just like that', in order to try out something new. He compared his current 'experiment' with starting rubber sixty years back; at that time, only a few farmers tried planting natural rubber, but nowadays rubber is the most important cash crop in the region. So it could well be that vanilla will also one day become one of the most important cash crops. For this reason, he wanted to test the crop now, but without having any real expectations from it.

Another institution active in the promotion of vanilla is the Vanilla Farmers' Society promoted by the cooperative society. In 2004, the society had around thirty members, who exchanged experience and knowledge about the crop. A further objective of the group was to be able to get a society loan from the local Cooperative Bank. Some years earlier, the Cooperative Bank introduced a new loan scheme for societies, which requires, however, that these have a minimum of twenty farmers and that they are formally registered with the Panchayat office (section 7.2.4). Only with a letter of reference from the Panchayat can they then open an account at the bank. The loan is then distributed amongst the members and managed by the group. Thus, the society has a double function: firstly, it organizes farmers interested in similar crops and secondly, it enables them to access the loans.

In summary, the Vanilla Promotion Scheme is used mostly by small to medium farmers of Types III, IV and V. They have the financial capacity to buy the necessary (and, at the time of the interviews, very expensive) seedlings to have access to the subsidies. They also possess enough land without too much shade to plant vanilla. For marginal farmers, however, the plantation of vanilla involves risks which are far too high. Their focus is on cultivating food crops (mixed farming) and rubber as a well-established cash crop. Vanilla, especially due to its uncertain future price development, is a crop which requires a higher risk-bearing capacity. Also, marginal rubber holdings do not have suitable land with enough water.

### 7.2.4 Loans from Informal and Formal Institutions

As has been described earlier, there are a great variety of financial institutions and organizations, from formal banks to smaller saving groups linked to self-help groups. This section describes how these institutions are connected to rubber holders and how they either support or hinder the rubber holders' diversification of their livelihoods and sources of income.

For Type I marginal holdings and some of Type II, two types of loans are important: firstly, smaller (Rs. 500 to 20,000) and short-term (two months to one year) loans help to secure the livelihood of these families during the seasons characterized by lower yields and less job opportunities, and secondly, larger and longer-term loans are crucial for their to improve their livelihood situation in the long term, for example, to acquire land and construct houses. When one considers the former type (smaller, short-term) more closely, many of the marginal holdings—which depend primarily on income from rubber and tapping activities—need extra income to get through the period when rubber gives no yield. Thus, during the low-yield season, many take loans from the local Cooperative Bank to pay for food; these loans are repaid as soon as the first income from rubber flows into the household again.[26] It is important to recall here the role of the rubber dealers who provide regular advances to rubber holders. These advances are repaid in kind when the rubber holders deliver rubber sheets to the dealer's shop. Another option for getting informal advance payments does exist: rubber holders working as tappers on larger holdings can get an advance on their salary, which will be deducted at the end of the month. Many tappers use this option on a regular basis. Loans from cooperative banks (and, to a lesser extent, from private banks) are mainly used for life-cycle events (weddings, dowries, funerals and other ceremonies), as well as for unexpected 'emergencies' such as serious health problems. Some of these loans are so-called 'gold loans'.[27] The different informal loan schemes implemented by the self-help groups in the Panchayat showed themselves to be an interesting alternative to smaller loans. For example, some marginal rubber holders take regular loans from *chitties*; these loans are very helpful because of their flexible and non-bureaucratic character.

Bigger and longer-term loans (i.e. investment loans) made by both cooperative and commercial banks for a specific investment purpose also play an important role in marginal holdings. Many farmers said that they were able to construct (and improve) their houses, to buy land, to plant rubber trees and also to buy cows only because they could access such investment loans. However, the regulations from the banks are very restrictive and they will only give loans against the deposit of formal landownership papers. However, the organization of collateral, especially of the landownership papers, is not always easy (see also Geiser 2001: 35-6). Also, the amount of the loan is linked to the value of the land (collateral), which can be a constraint if the farm is too small. Nonetheless, taking one of these loans is almost the only way for these holdings to improve their livelihoods in the medium and long term.

Bigger rubber holdings of Types III, IV and V have easier access to loans because of their regular agricultural income and the availability of land assets which can be used as collateral. Regular income from a fixed job (e.g., a civil servant's job) also eases access to loans and all holdings of Type IV have at least one member in such a job. Bigger and longer-term loans (around Rs. 200,000 to 400,000, for between 2 and 4 years) are needed for investments in land and buildings. Loans are especially important in the cultivation of cash crops on leased land, something larger holders often practise. In fact, land-leasing contracts require a big portion of the land rental to be paid in advance to the formal landowner. This is often only possible with a loan. Furthermore, on many of the holdings bigger loans have been instrumental in permitting the holders to start up non-agricultural business activities. For example, one rubber holder got a loan of Rs. 200,000 to build a smokehouse and open a rubber shop in Eratuppetta. Another got a loan of Rs. 365,000 to buy a vehicle for his taxi business. Here, the role of private moneylenders is also crucial in permitting rubber holders to raise the necessary capital as security for the banks. Private moneylenders also assist farmers when it becomes necessary to 'refresh' a loan, which means to pay back an old loan in order to get a new one from the same or another institution.

As mentioned above (section 7.2.3), special group loans exist for farmers who are formal members of a Farmers' Society (e.g., the Vanilla Farmers' Society). For some years, these societies have been able to open an account in the Cooperative Bank in Thalanadu; if they so wish they can get a group loan, which is then distributed to individual farmers in the group depending on their needs. Though the final interest rate is slightly higher than for an individual loan, the advantage for the farmers is that they can access this loan without having to go through all the legal formalities, for example, without having to deposit collateral. Also, the group can assist a farmer member almost immediately in case of necessity. For the bank, the advantage lies in the lighter workload than if it had to attend to each farmer individually. Since the scheme was quite new at the time of the field study, its actual importance for individual rubber holders could not be assessed from the interviews.

In summary, access to formal and informal loans is of the utmost importance for all types of holdings. For the smaller ones, both informal and formal loans help to pay for food and necessities during seasons with limited income options. For larger holdings, longer-term loans enhance the capacity of rubber holders to invest in new agricultural—but also in non-agricultural—activities. The option of optimizing the uses of different types of loans for different types of needs and refreshing loans whenever necessary is exercised by many holders.

### 7.2.5 Social Networks

Social networks have been mentioned many times earlier in relation to not only initiating new income opportunities, but also supporting rubber holders' livelihoods, for example, smaller loans from friends, colleagues and neighbours. Especially in marginal Type I holdings and some in Type II, the possibility of getting informal and smaller loans within these networks turns is of great importance (Box 7.1). The advantage is that there is no need to show either formal documents or collateral. Often, this money is used to repay a bank loan, but sometimes also for living costs. However, in almost all cases, rubber holders only resort to loans from neighbours or colleagues at a later stage; they first make use of formal loans from banks and credit cooperatives. Another way of taking a loan from banks is to ask friends or colleagues to put up their own landownership papers as collateral. This has happened in some cases (Box 7.1).

Loans from friends or colleagues were also mentioned in relation to investments in agricultural businesses such as planting crops on leased

**Box 7.1: The Importance of a Strong Social Network**

Siddarth, a successful agricultural labourer, had a bad accident some years back. As a result, he had to stop working. Since the family was not able to cover the household expenses with the income from their small rubber plantation, they first raised a total of Rs. 40,000 in loans from various friends. Some friends asked for a small percentage as interest, while others decided to give the money without interest. Siddarth needed more money to pay for important surgery. He took a loan of Rs. 10,000 from the Cooperative Bank in Thalanadu against his land documents. To repay part of the loan to his friends and also to the bank in Thalanadu, he took another loan of Rs. 10,000 from the Cooperative Bank in Teekoy. However, for this second loan, a friend had to 'lend' his formal land documents to the bank otherwise Siddarth would not have got the loan. To be able to repay the loan from the Teekoy Bank (twelve instalments of Rs. 500 each, including interest), he asked some friends and neighbours to help him out each month, though he had only just repaid them some months earlier. Siddarth thus managed to keep all his lenders more or less satisfied. As they were aware of his tragic story, they were ready to help him out. However Siddarth feels that in a few years' time this strategy will no longer work.

land. For example, one rubber holder decided to lease an old rubber plantation for slaughter tapping, he was able to do so only because he had a colleague who was ready to invest his own money in the venture. The deal was that they would divide the profit from the plantation at the end of the leasing period. However, it is crucial to find suitable plots to be leased. Friends and colleagues in the wider neighbourhood are essential here since they can give notice as soon as they know that a plot is available. Many of the rubber holders who lease land said that they had got access to good land only because some friends or colleagues had told them about its availability. Thus, a strong social network is the base for all successful business activities.

During the interviews in Choovoor ward, an important institution based on a strong local social network came to light: the organization of voluntary and free school classes for members of the SC/ST community. As it happened, a retired teacher had opened a small private school and was organizing free daily classes for people from the SC/ST community who lived in his neighbourhood. The main aim of the classes was to prepare these people for the civil service exams, which open the doors to the SC/ST reserved jobs available in the state or central government. According to one interviewee, the chances of people who attend these classes passing an examination and getting a job in the government are much higher than of those who do not. This could be the reason why—as a staff member of the Rubber Board highlighted[28]—a comparatively higher percentage of rubber holders in Choovoor have civil service jobs.

Finally, it is important to recall the role of the *Kudumbasree*-initiated local women's self-help groups (section 7.1.2). Women often get information about available income-generating schemes via these groups. For example, one woman started a training course in cow-rearing because she had heard about this possibility in a self-help group meeting.

### 7.2.6 Social Status of Rubber Holders

The social status of a family was found to play an important role with respect to access to diversified income-generating activities. Some of the interviewed rubber holders said that the social status of families has restrictive or enabling effects, as the following example shows.

In one of the families, the income from rubber during the period of low prices (the 'rubber crisis') was not sufficient to cover the current household expenses. Thus, one household member planned to seek work as an agricultural labourer on neighbouring farms to bring in some additional income. However, members of his extended family did not allow him to do such work because of the low status connected with agricultural labourer work. The family argued that, if he were to do such work, the status of the whole family would be lowered. As a consequence, the family preferred to face the burden of insufficient income, rather than let him work as an agricultural labourer. Another example is of leasing land for cash-crop plantations. In families with a higher social status, household members are normally not allowed to lease a small plot of land. If they did this, they would have to work in the fields themselves; they would thus be considered to be agricultural workers and of low social status. To avoid this type of judgement, they have to lease land of at least three to four acres. They are then considered to be businessmen—a higher social status—investing in land-leasing activities and employing and supervising agricultural labourers. Though the line between the former and the latter practice is blurred, it is important to understand that a certain social status can be an impediment in the pursuit of certain activities. Interviews show that this way of thinking is especially prevalent in many of the rubber-holding families of higher social status, normally the better-off Types III, IV and V holders. In contrast, it is a less important factor in families of lower social status, since they generally have little to lose.

The family's social status is also important when it comes to the role of women with respect to income generation. In families of higher status, women are not allowed to work outside the holding, and within the holding they are normally expected to do household rather than agricultural work. In cases where a male member is—sometimes temporarily, sometimes for longer periods—not able to work because of injuries or other health problems, the family can suddenly find itself in a very difficult financial situation. It can be even worse if the profits from rubber are reduced to a minimum due to the need to hire external tappers. As a final resort, these households will depend on the help of their neighbours (section 7.2.5). The important role of social networks, especially of informal loans from neighbours, was confirmed by a study analysing the income-generating and expenditure strategies of female-headed rubber households[29] in central Kerala (Zollinger 2005: 63). Interestingly, in lower status marginal holdings, women have the option of earning an income from any activities they can find. For example, they are active in agriculture within their rubber holdings, but also work outside the holding as agricultural labourers, workers in factories or sometimes, they may even migrate temporarily to factories in other states (Box 6.3, section 6.3.2).

## 7.3. SUMMARY

### 7.3.1 Institutional and Organizational Setting

As field research shows, a multitude of institutions and organizations have an impact at a local level. These institutions and organizations, however, are not necessarily based locally. The case of rubber marketing institutions, by contrast, shows that rubber holders actively look for organizations and institutions outside the *local area*; for example, to get higher prices for their rubber sheets. In both of the wards that were analysed, the mobility patterns of the different holders are very diverse.

While holders from Choovoor are mostly oriented towards institutions based in Moonilavu centre (because of its proximity), holders from Maravikkallu ward go either to the centre of Thalanadu Panchayat (Thalanadu centre) or to Teekoy centre, which is a little further away but offers better services and prices (Fig. 7.1). It is, however, important to differentiate between the mobility pattern of marginal holders (Types I and II) and that of bigger holders (Types III, IV and V). In marginal holdings, household members tend to travel shorter distances to market their products (rubber sheets, but other agricultural products too), as well as less often for example, to access services of institutions such as the agricultural office (agricultural schemes). The time and transaction costs involved are often too high compared to the expected (low) financial returns. The situation is different in larger holdings, for which higher mobility can be rewarding; an example is transporting rubber sheets to a nearby town to sell. Lack of spatial mobility does hinder rubber holders from being able to access the different types of institutions and organizations. This is shown by the detailed analysis of the institutions and organizations by different type of holding.

### 7.3.2 Institutions and Organizations: Impact on Marginal Holdings (Types I and II)

To recall, marginal holdings are of two types: those with agriculture as the main income sector (Type I) and those where non-agricultural income is more important (Type II). Analysis of the Rubber Board replanting subsidies shows that most marginal holdings do not access these replanting subsidies. There are many reasons for this: some farmers are not interested in these subsidies because they do not gain any advantage from them; others are interested but cannot access them because of the cost involved; and a final group of holders was simply not aware of their existence. The reasons for the low impact of these subsidies are not only the technical set-up of the scheme (e.g., the need to cut down most other non-rubber trees), but also the costs of joining the scheme, which are often higher than the expected returns, particularly in the case of the most remote holdings. The situation is similar for other non-rubber agricultural schemes, as shown by the analysis of the Vanilla Promotion Scheme. In fact, few marginal holdings have the possibility of accessing this scheme mainly because of their lack of capital (expensive seedlings), their lack of land and the lack of water on their plots. These two agricultural schemes neither promote nor hinder income diversification, mainly because these holdings only seldom make use of them.

For marginal rubber holdings, other institutions and organizations as well as institutionalized practices are of greater importance. A good example of the impact of institutionalized practices is the marketing practice implemented with local rubber traders. The regular small cash advances from rubber dealers to farmers, for example, allow the latter to get through periods of lower income. On the other hand, farmers can also ask for payment of the rubber sheets to be delayed if they expect prices to go up in the near future. Through these practices, farmers can both increase and even out their monthly income. The most important institutions and organizations, however, are the financial institutions, especially those providing loans. Analysis shows that marginal farmers have access to most of them; the limitation, however, is the size of the loan given, which reflects the lack of available collateral. For marginal holdings, these loans help first of all to sustain the household livelihood during the seasons of lower income; secondly, they help marginal farmers to invest in new housing facilities and sometimes to add small new plots of land to their holdings. Thus, these loans are sometimes crucial for farmers to be able to 'speed up' the improvement of their livelihoods. Finally, studies show the importance of social networks and social status to marginal holders. The (few) social networks analysed are seen to be important to access loans, for example, through the availability of collateral within the neighbourhood network. These are often the last resort, especially in times of crisis. Analysis of the impact of the social status of marginal holdings on income and diversification shows that it is neither strongly positive nor very negative.

### 7.3.3 Institutions and Organizations: Impact on Bigger Holdings (Types III, IV and V)

The impact of the local institutional and organizational setting on holdings of Types III, IV and V is quite similar, but different from that on Types I and II described in the earlier section.

These holders often access the governmental subsidies given through the Rubber Board Replanting Scheme and the Vanilla Promotion Scheme. Analysis of the Rubber Board replanting scheme shows that almost all the largest rubber holders (Type V) do receive subsidies when replanting their trees; however, those of Types III and IV, though receptive, do carefully weigh the advantages and disadvantages. As it happens, many of them have experienced that the constraints related to subsidies—especially the need to cut down other trees—outweigh the benefits. However, all these larger holders benefit from the research and extension services provided by the Rubber Board field officer, especially if they follow the production practices promoted by the Board to the letter. Thus, the scheme increases the income from rubber but hinders agricultural diversification within the rubber plantations.

Analysis of the Panchayat's Vanilla Promotion Scheme shows that a good proportion of holders in these categories take the opportunity to plant subsidized vanilla; however, they never go beyond the maximum number of subsidized seedlings. The reasons for this are the unavailability of land with sufficient water and, above all, the investment capability of these holders. However, vanilla was mostly planted as an experiment which could prove economical in the future. During research, a link was also established between membership of the organization of vanilla planters—the Vanilla Farmers' Society—and the possibility of getting an agricultural loan from the local Cooperative Bank. The Vanilla Scheme thus promotes on-farm agricultural diversification, especially in the bigger holdings. Income-wise, the results of vanilla planting have been limited, both by a lack of good growing conditions (water) and by the fall in the price of vanilla.

Loans from financial institutions are as crucial for these bigger holdings as they are for marginal holdings. The main difference, however, is that larger holders can access substantial loans more easily, from both the Cooperative and private banks. This is due to greater collateral being available on these farms, in terms of both land availability and the regular income from civil service jobs on certain holdings (especially Type IV). Access to this capital has allowed some of the farms to invest in larger-scale ventures, such as starting a new non-agricultural business or entering the land-leasing business. Thus, it can be concluded that the local financial institutions and organizations support new income activities, and in many cases strengthen diversification processes, within both the agricultural and the non-agricultural sectors. Diversification is also supported by or linked to social networks. In some cases, family members and/or friends were crucial in giving advice and strategic information to larger holders.

There are some institutions and organizations which do not promote income generation or income diversification on larger holdings. Some examples are the institutionalized marketing practices mentioned above in relation to rubber dealers and marginal holdings. In fact, bigger holdings do not need to embark on these activities since they have enough cash to withstand periods of low yield; larger holders can also stock their rubber sheets for longer periods and sell them to bigger market players under better conditions. Finally, analysis of the social status of larger holders has shown at least one case where the institution of 'social status' has a negative impact on the income of a household. In general, however, bigger holdings have better opportunities to access institutions and organizations and draw benefits from them.

## NOTES

1. Traders in Teekoy are ready to pay higher prices to farmers due to the lower transport costs from Teekoy to Eratuppetta, the next largest market town for such produce.
2. A reminder of the definition of 'social relation': it is 'the social positioning of individuals and households within society' (Table 2.4).
3. The distribution of Rubber Board field officers is not according to Panchayats but by village. The Rubber Board field officer in Teekoy is responsible for planters in the four revenue villages of Teekoy, Moonilavu, Poonjar North, and Poonjar Middle. This corresponds more or less to the area of two Panchayats, Teekoy and Thalanadu (interview with Rubber Board field officer, 6 May 2004).
4. The Tribal Development Rubber Plantation Programme (TDRPP) was started in 1997 in some nearby Panchayats (Meladukom, Vellani, Kollani, and Mangombe). Around 90 tribal families with a total of around 25 ha participated and planted rubber. The Rubber Board provided planting material and reimbursed the farmers for their labour costs (planting, weeding, manuring) over a six-year period. The RB also gave all inputs for free (e.g., manure). After this phase the programme was discontinued as there were not enough new families interested.
5. The World Bank website lays out the main goals of the Rubber Project (1994-8) as follows: 'The project will finance an expansion and rehabilitation phase of India's rubber sub-sector. It includes assistance for new plantings, replantings and processing of rubber and rubber wood. The project will strengthen the Rubber Board's research, extension and training activities and support its on-going programme to enhance productivity of mature rubber through improved management inputs. Development programme for women and tribal beneficiaries and Non-Governmental Organization participation

will be supported. The project will benefit rubber growers and rubber processors and create new jobs in off-farm activities' (*Source:* World Bank, Rubber Project India, http://web.worldbank.org/external/projects/main?pagePK=64312881&piPK=64302848&theSitePK=40941&Projectid=P009959). Accessed on 9 September 2006.

6. Rubber Board, Housing and Sanitary Scheme for SC/ST Rubber Tappers, http://rubberboard.org.in/labourschemes.asp?id=13 Accessed on 9 September 2006.
7. Interview with Rubber Board representative, Kottayam, 8 March 2004.
8. Some of the schemes sponsored by the central government are implemented via the state government agricultural department, others via the Commodity Boards, such as the Spices Board or the Coconut Board. In the field, the agricultural officer is responsible for both. For example, the Spices Board has made it the responsibility of the local agricultural officer to implement the scheme to promote the plantation of pepper (subsidies for seedlings and manure). The Rubber Board is an exception to the rule on implementation, since it has its own officer in the field, who has complete responsibility for issues concerning natural rubber.
9. Interview with the agricultural officer, Thalanadu, 5 March 2004, and the agricultural assistant, Thalanadu, 17 May 2004.
10. Interview with a spice trader, Thalanadu village, 11 May 2004.
11. According to the spice trader, the price of nutmeg in May 2004 averaged Rs. 175 per kg, whereas it had previously been around Rs. 35 per kg.
12. Interestingly, interviewees mentioned that they went to moneylenders when bank loans needed to be 'refreshed'—which means paid back—so that they could get a new loan from the same or a different bank.
13. The income limit under which a family is defined as BPL is defined regularly by the central government. In 2003-4, a BPL family was defined as having an income of less than Rs. 325 per month. However, according to a Panchayat member, this definition is more relevant to income levels in north India. For Kerala, it has to be adapted. In Thalanadu Panchayat, the *Grama Sabha* makes a list of BPL people who are eligible for the schemes (interview with Panchayat member, Thalanadu, 23 January 2004).
14. The following example shows the difficulty in getting a civil service job: in 2004, the examination for one of the 150 vacant posts as a government forest guard was taken by 20,000 candidates. Of the 150 vacant posts, about 20 are reserved for SC people, and 20-30 for ST people. According to the interviewee, about 1,500 SC people sat for the examination. Thus, for non-SC/ST people, the chance of getting a post is 0.56 per cent, for a ST member it is 1.33 per cent.
15. A *chitty* is a special type of rotating saving and credit association.
16. Onam is the biggest annual festival of Kerala.

17. Interview with the Rubber Board field officer, Thalanadu, 06 May 2004.
18. The reference price would be based on a 7-year moving average of international prices. When the price falls below 20 per cent of this reference price, growers could benefit from the relief fund. But when the price goes above 20 per cent, the growers have to contribute to the fund (*Business Line* 2003).
19. According to the local field officer, a maximum of 40 coconut trees or 80 arecanut trees are allowed.
20. As an example, one farmer having 0.62 acres (0.25 ha) of land got Rs. 400 the first year, Rs. 350 the second year, Rs. 280 the third year, Rs. 135 the fourth year, and even less in the last two years.
21. Some of the interviewed farmers said that this seemed to be a new (informal) practice of the Rubber Board field officer.
22. It is clear that one must be careful in doing this calculation: for the sake of exactitude, the value of these timber trees in the 1990s and the inflation rate should be included in the calculation.
23. In fact, it would be too complicated for an RPS manager to deal with the many daily changes in the rubber price. This can be both an advantage and a disadvantage.
24. Interview with a Panchayat member of Maravikkallu ward, Thalanadu, 23 January 2004.
25. Interview with the agricultural officer, Thalanadu, 5 March 2004.
26. During 2003-4, the interest rate on these loans averaged 10.5 per cent per annum.
27. See section 6.6.2.
28. Discussion with an expert of the Rubber Research Institute of India, Kottayam, 9 March 2005.
29. Female-headed households can be categorized in de jure and de facto headships. De jure female-headed households are those in which a woman is registered as the legal head. This usually happens when there is no male head. Such women are unmarried, divorced, widowed or deserted. A de facto female-headed household is defined by a situation where a nominal male head is present, but the woman bears the main responsibility of the household (Zollinger 2005: 8).

CHAPTER 8

# Conclusions and Implications

Although many countries have been linked to global agricultural commodity markets for hundreds of years, the last two decades have seen a huge wave of globalization and economic integration. Since the beginning of these processes of globalization, deregulation and economic liberalization, research studies have been showing that the uncertainty affecting agricultural markets and the opening of rural areas to the 'global world' have brought with them a shift in smallholder income from solely agricultural and farm income towards a more diverse income portfolio. The present study takes this as its research hypothesis and seeks to validate it through a crop- and locality-specific case study. It differentiates between *diversity*, i.e. the diverse status of the holding at a specific moment, and *diversification*, or the process of change from one portfolio of activities to another one within a specific time frame (section 2.2.2).

Section 8.1 first recalls the typology of rubber holdings used in the study. Then, it summarizes the main overall findings of the study on the diversity and diversification of natural rubber holdings in Kerala, and on the role of the local organizations and institutions. Section 8.2 presents some conceptual and theoretical considerations. Finally, section 8.3 highlights a number of policy implications.

## 8.1 MAIN FINDINGS

### 8.1.1 TYPOLOGY OF NATURAL RUBBER HOLDINGS

Analysis has shown that it is necessary to differentiate between different types of smallholdings to be able to come to any meaningful conclusions about their livelihood situations. In fact, multiple livelihood portfolios exist even within the same locality and context. Thus, a typology of natural rubber holdings is developed in this work to allow for a better understanding and sharper analysis of the income-generating strategies of rubber holders. The definition of the size of a rubber holding as used in this study differs from the Rubber Board's official definition. Here, the size of a rubber holding is taken to be the total holding size (i.e. all land belonging to the members of a household) and not just the total area under natural rubber. The rationale behind this choice is the fact that when studying the livelihoods of farmers all relevant income sources have to be included in a study. Indeed, the size of a rubber plantation was also defined differently than suggested by the Rubber Board. In this study, the total size of the rubber plantation is defined as the total number of rubber trees planted on the holding land (and not in terms of land area as is the case with the Rubber Board). This choice was made for three reasons: firstly, the total number of rubber trees takes into account the differences in tree densities which often occur in marginal and hilly areas; secondly, the frequent loss of rubber trees due to storms or disease can be taken into account; and thirdly, it is normal practice for rubber holders to speak about their rubber plantations in terms of total trees and not area.

The typology distinguishes between five types of holdings is shown in Table 8.1. Types I and II consist of marginal holdings with an average total land area of 0.51 ha and 0.29 ha respectively, and 88 and 51 rubber trees respectively. While Type I holders have multiple strategies with different activities in the agricultural and non-agricultural sector on which they depend for their basic livelihood, Type II holders are far better off, as they earn a decent income from being either salaried employees or

TABLE 8.1: TYPOLOGY OF RUBBER HOLDINGS

| *Type* | *Size category* | *Main income sector* | *Yearly gross income* | *Average total land (ha)* | *Average total rubber trees* |
|---|---|---|---|---|---|
| I | Marginal (<1 ha) | agric. | low | 0.51 | 88 |
| II | Marginal (<1 ha) | non-agric. | low | 0.29 | 51 |
| III | Small (1-2 ha) | agric. | medium | 1.40 | 316 |
| IV | Small (1-2 ha) and Semi-medium (2-4 ha) | non-agric. | good | 1.59 | 366 |
| V | Semi-medium (2-4 ha) and Medium (4-10 ha) | agric. | very good | 3.41 | 971 |

*Source*: Author's survey and interviews; for details see section 6.2.

self-employed. Holdings in Types III and IV are small to semi-medium with around 1.40 ha and 1.59 ha of land respectively, and the number of rubber trees owned is 316 and 366 respectively. While Type III holdings focus exclusively on rubber production and earn their income from it, Type IV holders derive their main income from a non-agricultural activity whilst employing labourers to take care of their rubber. The last type of holdings represented in the case study area (Type V) encompasses semi-medium and medium holdings with an average of 3.41 ha of land and 971 rubber trees (section 6.2).

### 8.1.2 Role of Diversified Livelihood and Diversification among Rubber Holdings

Analysis of the income-generating activities of different types of rubber holders in Thalanadu shows that they base their livelihoods on a heterogeneous portfolio of activities. The extent of the diversified status and the type of income activities followed depends on different factors, one of the most important being the total size of the holdings. The most diverse portfolios are found at the extremes of the size range of the holdings, in the most marginal (Types I and II) as well as in the biggest holdings (Types IV and V), which also corresponds to the poorest and the richest ones. The holdings in the middle of the range (Type III) are less diverse and more specialized in rubber cultivation.

The poorest and most marginal holdings derive only a small percentage of their income from their own rubber trees; however, they earn important income from their work as tappers and labourers on neighbouring plantations and are thus altogether very dependent on income from the natural rubber sector. Mixed crops including staple food crops play an important role in the daily subsistence of these holdings. The combination of mixed food crops with rubber trees provides sub optimal results usually in terms of yield; however, the combination is important to reduce the dependency on one or a few crops. Income from the non-agricultural sector, i.e. the manufacturing and services sectors, plays a key role for certain holdings, at least seasonally. For marginal holdings, the key driving force towards a diverse income situation is found to be their vulnerability due to their limited monetary income. Other factors such as seasonality, climatic uncertainties and price fluctuations also influence their decision-making, which can best be described as risk-minimizing. The greater overall geographical marginality of these holdings—with longer distances to roads, as well as steeper and hillier plots—also has a negative impact on their cropping options and their access to services and institutions.

The largest holdings also have a diverse income portfolio, some favouring agricultural activities (cash crops on leased land) and others non-agricultural ones (especially civil service jobs). The driving forces on these holdings are however substantially different from those on marginal holdings. The farmer tend to have diverse income portfolios because of the opportunities they have to access new activities: investment possibilities due to their access to financial resources; the availability of land resources; and especially their ability to take risks and cope with failure if necessary. Medium-size holdings do not show such high diversity; they tend to specialize in natural rubber. This is more than just a choice —it is the size of their rubber plantations that forces them to focus on the production of and income from rubber. In fact, the size of these holdings requires them to dedicate most of their household labour resources to tapping activities, since employing external tappers is too expensive. And yet they are still too small to be able to take the risk of becoming involved in other business activities, as the biggest holdings do, for example. Although these holdings manage to cope well with just the income from rubber, they are the ones that are most dependent on the fluctuations in the price of natural rubber.

The main findings about the diversification process—i.e. the changes from one portfolio of activities to another one between 1995-6 and 2003-4—are the following. As seen earlier, diversity is high in Types I and II holdings since they cannot make a living from rubber on its own and need to be (and remain) diversified in order to cope. Since these holdings already displayed some diversity in 1995-6, their diversification process was found to be low to medium, though in some cases the 'rubber crisis' had slightly accentuated the need for further diversification. The situation is very similar for Type IV holdings (small and semi-medium non-agricultural holdings). As they were already pursuing most of their non-agricultural activities even before the 'rubber crisis', no process of diversification has taken place over recent years. For holdings of Types III and V, it can be concluded that the process of diversification is higher than on other holdings; it has been assessed as being medium. On Type III holdings this is because some young planters have started business ventures in recent years; their strategy of becoming less dependent on agricultural income is also complemented by the possibilities they now have to invest in new activities. Type V holdings have also responded to opportunities to invest in new activities, especially the cultivation of cash crops on leased land. As these farmers started leasing activities only recently, their diversification process is more accentuated than in other holdings—but it is an agricultural diversification.

### 8.1.3 Role of the Local Institutional and Organizational Setting in Promoting or Hindering Diversification

The second hypothesis of the study states that the local institutional setting plays a key role in hindering or supporting the diversified livelihood strategies of natural rubber holders. A range of local institutions and organizations was studied for this work, namely, the Rubber Board replantation subsidies, the rubber marketing institutions, the Vanilla Promotion Scheme, loans from informal and formal institutions, social networks and social status. The research for this study shows that most play an important role in promoting or hindering diversified incomes. However, a first conclusion is that each institution affects each type of holding in a different way. Institutions can have a promotional effect of diversified livelihoods on some types of holdings, whilst being restrictive for others. This demonstrates the need for differentiated analysis.

It is the bigger holdings that make the greatest use of the agricultural schemes (Rubber Board replantation subsidies and vanilla subsidies). They have the resources to handle the necessary administrative work to participate in the schemes. By contrast, marginal holdings are either not informed about these schemes or not interested; the costs involved in getting the subsidies (travel, time) are too high compared to the low expected returns due to the small size of their plots. However, other constraints can be noted for marginal holdings: the Rubber Board replanting scheme is badly adapted to the mixed cropping habits of these holdings; and the vanilla scheme is not well-suited due to the 'marginal holdings' lack of investment possibilities and of land suitable for planting vanilla.

The impact of these schemes on diversification takes two main forms. If farmers can access it, the vanilla scheme promotes diversification into a new cash crop within agriculture. The present Rubber Board scheme, on the other hand, does hinder agricultural diversification because the rules do not permit other crops to be planted within the subsidized rubber plot. Thus, the scope of agricultural schemes to support diversification depends to a great extent on their set-up, but also on their accessibility by the different types of holdings.

The main findings on the subject of loans by informal and formal institutions are set out below. Analysis of these institutions shows that they are of crucial importance to all types of holdings. Most holdings manage to access most loans: a limitation for marginal holdings is that they can only take out small loans due to their lack of collateral. For these holdings, loans are first of all to support the household during seasons with lower incomes. Their second function is to help marginal farmers invest in new housing facilities and, in a few cases, to add small new plots to their holdings. Thus, in marginal holdings, the role of loans with respect to promoting diversification is limited. The situation is different in the largest holdings. Here, the most important function of loans is to start new income-generating activities, such as the production of cash crops on leased land. The conclusion is therefore that, on these types of holdings, loans promote diversification, sometimes within agriculture, sometimes into sectors other than agriculture.

The other institutions and organizations that were analysed, namely, the rubber marketing practices, social networks and social status, show no demonstrable promoting or hindering effect. They are of greater importance for the more vulnerable marginal holdings to be able to cope in times of crisis. Only 'social status' proved to hinder some farmers on larger holdings from accessing lower status jobs — and extra income —during the crisis. Their families prevented them from accepting these jobs.

## 8.2 CONCEPTUAL AND THEORETICAL CONSIDERATIONS

In this work, *livelihood diversification* has been defined in Ellis' terms (1998: 4) as 'the process by which rural families construct a diverse portfolio of activities and social support capabilities in their struggle for survival and in order to improve their standards of living' (section 2.2.2). The present work shows that Ellis' broad definition comprising, for example, social institutions, gender relations and property rights is appropriate for a study of livelihood diversification. In fact, when looking closely at the determinants of a process of diversification, it is important to note that the important factors are institutional support systems, social status and intra-household gender relations, as well as the opportunity and wish for social mobility. Thus, the present study is in full agreement with scholars who state that livelihood diversification is far more than just income diversification.

This was my point of departure when selecting a different analytical approach from that used in economic studies of diversification, which mainly consider sources of income. This work has developed a typology of rubber holdings which included quantitative income data from household activities alongside qualitative data: the reasons and motives of household members for diversifying (or specializing) and the rationale behind their diverse (or specialized) portfolio of activities. The final typology presented is therefore a combination of qualitative and quantitative factors (section 6.2.2).

My experience with this approach leads to two main conclusions. On the one hand, I am of the opinion that the typology with the five types of holdings proposed is applicable to most rubber holdings in central Kerala (excluding the large plantations, which were not researched for this study) and therefore of interest to other studies of the Indian smallholding rubber sector. The typology also presents the case for looking at and understanding each type of holdings separately. On the other hand, the blend of qualitative and quantitative approaches makes the typology difficult to grasp if one does not read the steps of the typology development with care. A further disadvantage of working with a typology is that holdings, once put in a typology, seem to be permanently fixed in that typology. This, however, is not necessarily the case, as holdings can be affected by downward or upward social mobility. In this work, mobility between types of holding could not be analysed in any detail because holdings were distinguished with regard to their situation in 2003-4. However, an assessment was made of the mobility pattern (upward *vs.* downward mobility) and of the determinants for social mobility (section 6.8.3).

Ellis' critique that many studies that aim at studying diversification are actually looking at diversity instead has been taken seriously in this piece of research. During analysis, a clear separation in nomenclature was established between diversification (as a *process*) and diversity (as a *status*). The consequence of this for the fieldwork was that the status of diversity had to be studied at two different moments and then compared. As the results of this study show, this differentiation was crucial to making a realistic assessment of the role of diversification over the period 1995-6 to 2003-4. If only the situation in 2003-4 had been examined, some of the holdings that were already quite diverse in 1995-6 but had not changed much since then would have been labelled as 'having diversified' whereas in reality they did not diversify but remained diverse. Understanding this makes one question the many studies that claim to study diversification when what they are actually showing is the status of diversity at a given time.

The strict classification of assets, activities and incomes into wage- *vs.* self-employment, agricultural *vs.* non-agricultural employment and on-farm and off-farm as suggested by Barrett et al. (2001: 319) also turned out to be a good choice. By putting income into these categories, it became possible to compare own data with data from other studies using the same criteria, for example, distinction between sectors. In some cases, a further subdivision of the spatial classification 'off-farm' into 'local' (within the village or region), 'domestic' (outside the region but in the country) and 'foreign' (outside the country) would have highlighted the role of periodic national and international migration. To some extent, this was done in another study (Bichsel et al. 2005).

As already mentioned, studies on the diversification of livelihoods started to appear in the 1990s, but the bulk of the literature on different aspects of diversification has only come out in recent years (section 2.2.1). Certainly these publications have corrected the view that most rural households depend solely on agricultural income; they clearly state that, in many countries, rural people derive around half their income from other sources. Though this might well be right on average, the present study shows that it is necessary to investigate what lies behind averages and also focus on the heterogeneity of income portfolios which exist in rural areas. Only by understanding the differences between the situations of the many households adequately can policy measures be taken. For example, the case study presented here contradicts the general hypothesis that the changes during recent years, i.e. deregulation and economic liberalization, are instrumental in the transition to more diversified livelihood strategies. Again, differentiated analysis is required. In fact, most smaller holdings were already diversified in the mid-1990s, before the major impact of deregulation and economic liberalization; even the 'rubber crisis' that affected rubber growers in the second half of the 1990s did not much change their situation. Instead of diversifying their income activities, many holdings reduced household expenditure. Only a few larger holdings diversified into new income-generating activities in the non-agricultural sector; the driving factor for them was the opportunities rather than the necessities.

A statement from Ellis provides an appropriate conclusion: 'the causes and consequences of diversification are differentiated in practice by location, assets, income, opportunity and social relations; and . . . these manifest themselves in different ways under different circumstances' (Ellis 1998: 3). This is a plea for researchers to adopt a differentiated approach and to put livelihood diversification studies into a specific context.

## 8.3. POLICY IMPLICATIONS

It is suggested here that the main findings of this case study have the following implications for policy.

A first implication is that the expressions 'rubber holding' and 'rubber small holding'—as generally used in literature on the Indian rubber sector—are quite vague and do not pay sufficient attention to the fact that these holdings are a heterogeneous group with diverse and specific livelihood characteristics. It is fundamental to detail which types of holdings one means when using these expressions. Secondly, the extremely heterogeneous situation of rubber holdings implies that any institutional support wishing to improve the livelihood of rubber growers has to be adapted to the different situations and the different needs of each type of holding. This is especially true in the case of institutions which support agricultural production in general and rubber production in particular. In truth, the different types of holdings have rubber management practices that are adapted to their socio-economic and geographical situation. While the most marginal holdings do not normally follow the scientific extension package of the Rubber Board, the bigger holdings generally do. It is thus important to acknowledge that many (mostly marginal) rubber holdings have difficulties accessing the rubber plantation schemes and support mechanisms, with the exception of some of the smaller schemes that specifically focus on these holdings. This lack of access to support mechanisms can be seen as an important constraint to the development of livelihood opportunities within the rubber sector, especially if one looks at the high proportion of marginal holdings in Kerala. In fact, 1985-6 data from the Agricultural Census shows that operational landholdings with less than 1 ha (the marginal ones) account for 92 per cent of all holdings in Kerala (Farm Information Bureau, 1995, in Véron 1999: 96).

The present study suggests that combining different agricultural and non-agricultural incomes is a widespread and significant practice, especially within the marginal holdings. It must therefore be considered a failing that there is no institution helping holdings with comprehensive long-term planning and implementation of all their on-farm and non-farm activities. Farm holdings have to deal instead with a multitude of sectoral schemes, each focusing, for example, on a single crop and often unrelated to each other. This fact is also highlighted by Barrett et al. (2001: 327). They state that the current challenges stem from the fact that issues concerning the rural non-farm economy are rarely backed up by any research and policymaking institutions. Agricultural researchers and policy institutions do not believe that non-farm activities fall within their mandate. And because many of the non-farm activities are rural, informal and generally small-scale means that the area is neglected by those involved in industry and employment policies and research; they prefer to work with urban and medium- to large-scale enterprises in the formal economy.

As Christopher Delgado and Ammar Siamwalla (1997: 2) state, there seems to be a gulf between the way policymakers and economists look at diversification. While policymakers view diversification as a major objective of rural development, economists see farm diversification more as an outcome of pursuing other objectives. Economists therefore look at diversification as the endogenous outcome of macroeconomic policies affecting input or output factor prices (e.g., fertilizer subsidies). Given the importance of non-agricultural income in Kerala, the time has come for a debate which would help policymakers to decide whether diversification and diversity should become a clear objective of rural development in the rubber holding communities in the years to come. If this is to be the case, institutional support has to be intelligently devised and tailored to the support of the different types of holdings in their effort to diversify their livelihood.

# Bibliography

Abdulai, Awudu and Anna Crole-Rees. 2001. 'Determinants of Income Diversification Amongst Rural Households in Southern Mali'. *Food Policy* 26: 437-52.

Alderman, Harold and Christina H. Paxson. 1992. 'Do the Poor Insure? A Synthesis of the Literature on Risk and Consumption in Developing Countries'. Policy Research Working Paper No. 1008. Washington DC: The World Bank.

Anderson, Edward, and Priya Deshingkar. 2005. 'Livelihood Diversification in Rural Andhra Pradesh, India'. *Rural Livelihoods and Poverty Reduction Policies*, ed. Frank Ellis and H. Ade Freeman. London: Routledge, pp. 62-81.

Anderson, Jock R. 2003. 'Risk in Rural Development: Challenges for Managers and Policy Makers'. *Agricultural Systems* 75: 161-97.

Appendini, Kirsten, Monique Nuijten and Vikas Rawal. 1999. *Rural Household Income Strategies and Interactions with the Local Institutional Environment: A Methodological Framework.* Rural Institutions and Participation Service (SDAR). FAO Rural Development Division. Available from: http://www.fao.org/sd/rodirect/roan0021.htm, accessed on August 2005.

Ashley, Caroline and Simon Maxwell. 2001. 'Rethinking Rural Development'. *Development Policy Review* 19: 395-425.

Ashley, Caroline, Daniel Start, Rachel Slater and Priya Deshingkar. 2003. *Understanding Livelihoods in Rural India: Diversity, Change and Exclusion.* Overseas Development Institute. Available from: http://www.livelihoodoptions.info/papers/guidance.htm, accessed on 12 August 2005.

Baak, Eric Paul. 1997. *Plantation Production and Political Power. Plantation Development in a Long-Term Historical Perspective, 1743-1963.* New Delhi: Oxford University Press.

Backhaus, Norman. 2001. 'Ökotourismus in malaysischen Nationalparks - Methodentriangulation in der sozialgeographischen Asienforschung'. *Asiatische Studien - Études Asiatiques* 4: 943-51.

——— 2003. 'The Globalisation Discourse'. IP6 Working Paper No. 2. Zurich: Development Study Group.

Backhaus, Norman, and Myriam Steinemann. 2001. *Leitfaden für wissenschaftliches Arbeiten*. Zürich: Geographisches Institut, Abteilung Anthropogeographie.

Barlow, Colin, Sisira Jayasuriya and Tan Suan. 1994. *The World Rubber Industry*. London: Routledge.

Barrett, Christopher B., Thomas Reardon, and Patrick Webb. 2001. 'Non-farm Income Diversification and Household Livelihood Strategies in Rural Africa: Concepts, Dynamics, and Policy Implications'. *Food Policy* 26: 315-31.

Bichsel, Christine, Silvia Hostettler and Balz Strasser. 2005. 'Should I Buy a Cow or a TV?' *Reflections on the Conceptual Framework of the NCCR North-South based on a Comparative Study of International Labour Migration in Mexico, India and Kyrgyzstan*. NCCR North-South dialogue. Berne: NCCR North-South.

Binswanger, Hans and Klaus Deininger. 1997. 'Explaining Agricultural and Agrarian Policies in Developing Countries'. *Journal of Economic Literature* 35: 1958-2005.

Binswanger, Hans and Ernst Lutz. 2000. 'Agricultural Trade Barriers, Trade Negotiations, and the Interests of Developing Countries'. Paper presented at the International Association of Agricultural Economists Meeting in Berlin, August 2000: Rural Development and Environment Department for Africa, World Bank.

Boeke, J.H. 1953. *Economics and Economic Policy of Dual Societies as Exemplified by Indonesia*. New York: Institute of Pacific Relations.

Booth, David. 1994. 'Rethinking Social Development: An Overview,' in *Rethinking Social Development: Theory, Research and Practice*, ed. David Booth. Essex: Longman Scientific & Technical, pp. 3-34.

Bruinsma, Jelle (ed.), 2003. *World Agriculture: Towards 2015/2030: A FAO Perspective*. London: Earthscan Publications.

Brutschin, Jeannine. 2002. *A Brief Review of Conceptualisations and Uses of 'Vulnerability'*. An Input for the Scientific Forum, NCCR North-South. Berne: Centre for Development and Environment.

Bryceson, Deborah Fahy. 1996. 'Deagrarianization and Rural Employment in Sub-Saharan Africa: A Sectoral Perspective'. *World Development* 24: 97-111.

——— 1999. 'African Rural Labour, Income Diversification and Livelihood Approaches: A Long-term Development Perspective'. *Review of African Political Economy* 26: 171-89.

——— 2000a. 'Peasant Theories and Smallholder Policies: Past and Present', in *Disappearing peasantries? Rural labour in Africa, Asia and Latin America*, ed. Deborah Fahy Bryceson, Cristobál Kay and Jos Mooij. London: Intermediate Technology Publications, pp. 1-36.

——— 2000b. 'Rural Africa at the Crossroads: Livelihood Practices and Policies'. *Natural Resource Perspectives*.

Burger, Kees, V. Haridasan, R.G. Unny, Hidde P. Smit and Wouter Zant. 1995. *The Indian Rubber Economy: History, Analysis and Policy Perspectives.* New Delhi: Manohar.

Burger, Kees and Hidde P. Smit. 2001. 'Economic Growth and the Future of Natural Rubber'. Paper presented at International Conference on the Future of Perennial Crops, 26 October 2001, Yamoussoukro, Côte d'Ivoire.

*Business Line.* 2003. 'Price Stabilisation Fund May not Benefit All', *The Hindu Business Line,* 29 March 2003.

Byerlee, Derek, Larry Harrington, and Donald L. Winkelmann. 1982. 'Farming Systems Research: Issues in Research Strategy and Technology Design'. *American Journal of Agricultural Economics* 64: 897-904.

Carney, D. (ed.), 1998. *Sustainable Rural Livelihoods: What Contribution Can We Make?* London: Department of International Development.

Centre for Earth Science Studies. 1984. *Resource Atlas of Kerala.* Trivandrum: Government of India.

Chakraborty, Achin. 2005. 'Kerala's Changing Development Narratives'. *Economic and Political Weekly*: 541-7.

Chambers, Robert. 1980. *Rapid Rural Appraisal: Rationale and Repertoire.* Sussex: Institute of Development Studies. Discussion Paper 155.

Colatei, Diego, and Barbara Harriss-White. 2001. 'Rural Credit', in *Reforms and Development: Essays on Long Term Village Change and Recent Development Policy in South India,* ed. Barbara Harriss-White and S. Janakarajan, London: DFID, Chaps 2-5. Available from http://www.livelihoodoptions.info/resources/change&policy.htm, accessed on 21 August 2002.

Corbett, Jane. 1988. 'Famine and Household Coping Strategies'. *World Development* 16: 1099-1112.

Crole-Rees, Anna. 2002. 'Rural Household Strategies in Southern Mali. Determinants and Contribution of Income Diversification to Income Level and Distribution'. PhD thesis No. 14,596. Zurich, Swiss Federal Institute of Technology (ETHZ).

Davies, Susanna. 1996. *Adaptable Livelihoods: Coping with Food Insecurity in the Malian Sahel.* Basingstoke: Macmillan Press.

Davis, Benjamin, Thomas Reardon, Kostas Stamoulis, and Paul Winters (eds.), 2002. *Promoting Farm/Non-farm Linkages for Rural Development—Case Studies from Africa and Latin America.* Rome: Food and Agriculture Organization.

Davis, Junior R. 2004. *The Rural Non-Farm Economy, Livelihoods and their Diversification: Issues and Options.* Chatham, UK: Natural Resources Institute.

Davis, Junior R., and Dirk Bezemer. 2004. *The Development of the Rural Non-Farm Economy in Developing Countries and Transition Economies: Key Emerging and Conceptual Issues.* Chatham, UK: Natural Resources Institute.

de Haan, Leo J. 2000. 'Globalization, Localization and Sustainable Livelihood'. *Sociologia Ruralis* 40: 339-65.

de Haan, Job, Gerard Groot (de), Egon Loa and Mark Ypenburg. 2003. 'Flows of Goods or Supply Chains: Lessons from the Natural Rubber Industry in Kerala, India'. *International Journal Production Economics* 81-2: 185-94.

de Haan, Leo, and Annelies Zoomers. 2005. 'Exploring the Frontier of Livelihoods Research'. *Development and Change* 36: 27-47.

De Soto, Hernando. 1992. *Marktwirtschaft von unten. Die unsichtbare Revolution in Entwicklungsländern.* Zürich und Köln: Orell Füssli.

Delgado, L. Christopher, and Ammar Siamwalla. 1997. *Rural Economy and Farm Income Diversification in Developing Countries.* Washington DC: IFPRI (Markets and Structural Studies Division).

Development Study Group. 2001. *NCCR North-South, IP6 Institutional Change and Livelihood Strategies. Phase Plan (07/01 - 06/05).* Zurich: Development Study Group, University of Zurich.

Devereux, Stephen. 2001. 'Livelihood Insecurity and Social Protection: A Re-emerging Issue in Rural Development'. *Development Policy Review* 19: 507-19.

DFID and John Farrington. 2004. 'Recognising and Addressing Risk and Vulnerability Constraints to Pro-poor Agricultural Growth'. Paper towards the Development of New Guidelines for Agricultural Policy in DFID. London: DFID.

DFID, Ian Gillson, Steve Wiggins and Nilah Pandian. 2004. 'Rethinking Tropical Agricultural Commodities'. Paper towards the Development of New Guidelines for Agricultural Policy in DFID. London: DFID.

Directorate of Economics and Statistics, Government of Kerala. Various issues. *Agricultural Statistics in Kerala.* Trivandrum.

Dixon, John, Aida Gulliver, and David Gibbon. 2001. *Farming Systems and Poverty. Improving Farmers' Livelihoods in a Changing World.* Rome and Washington: FAO and World Bank.

Douglas, Mary. 1987. *How Institutions Think.* London: Routledge & Kegan Paul.

Ellis, Frank. 1998. 'Household Strategies and Rural Livelihood Diversification'. *Journal of Development Studies* 35: 1-38.

——— 1999. 'Rural Livelihood Diversity in Developing Countries: Evidence and Policy Implications'. *Natural Resource Perspectives (ODI)*: April 1999.

——— 2000a. 'The Determinants of Rural Livelihood Diversification in Developing Countries'. *Journal of Agricultural Economics* 51: 289-302.

——— 2000b. *Rural Livelihoods and Diversity in Developing Countries.* Oxford: Oxford University Press.

Ellis, Frank, and Stephen Biggs. 2001. 'Evolving Themes in Rural Development 1950s-2000s'. *Development Policy Review* 19: 437-48.

Ensminger, Jean. 1992. *Making a Market. The Institutional Transformation of an African Society*. Cambridge: Cambridge University Press.

EPW Editorial. 2001a. 'Rubber: Looking Inward'. *Economic and Political Weekly* 2002.

——— 2001b. 'Rubber: Lost Bounce'. *Economic and Political Weekly* 2002.

——— 2004. 'Rubber: On the Rebound'. *Economic and Political Weekly* 2004.

Flick, Uwe. 1998. *An Introduction to Qualitative Research*. London: Sage Publications.

Flyvberg, Bent. 2004. 'Five Misunderstandings about Case-Study Research', in *Qualitative Research Practice*, ed. Clive Seale, Gianpietro Gobo, Jaber F. Gubrium and David Silverman. London and Thousand Oaks, CA: Sage Publications, pp. 420-34.

Franke, Richard W. and Barbara H. Chasin. 1998. 'The Kerala Model of Development: A Debate (Part 1)'. *Bulletin of Concerned Asian Scholars* 30: 25-36.

Geiser, Urs. 2001. 'To "Participate" with Whom, for What (and against Whom): Forest Fringe Management along the Western Ghats in Southern Kerala', in *Analytical Issues in Participatory Natural Resource Management*, ed. Vira Bhaskar and Roger Jeffery. New York: Palgrave, pp. 19-36.

Geiser, Urs, Andreas Tarnutzer, and Samuel Wälty. 1996. *Land Use Dynamics in Kerala: Towards an Actor-oriented Approach*. Zurich: Department of Geography, University of Zurich.

George, K. Tharian. 1999. 'The Natural Rubber Sector: Emerging Issues in the 1990s', in *Kerala's Economic Development*, ed. B.A. Prakash. New Delhi: Sage Publications, pp. 186-99.

George, K. Tharian, V. Haridasan and B. Sreekumar. 1988. 'Role of Government and Structural Changes in Rubber Plantation Industry'. *Economic and Political Weekly* 158-66.

George, K. Tharian, Joby Joseph and Jom Jacob. 2003. 'Global Trade and Tariff Policy on Rubber and Rubber Products Under the WTO Regime. A Preliminary Assessment'. Working Paper No. ER/1. Kottayam: Rubber Research Institute of India.

George, K. Tharian, Toms Joseph and Joby Joseph. 2002. 'Natural Rubber in Post-QRs Regime'. *Economic and Political Weekly* 37: 3319-21.

George, K. Tharian and Joby Joseph. 2005. 'Value Addition or Value Acquisition? Travails of the Plantation Sector in the Era of Globalisation'. *Economic and Political Weekly*: 2681-7.

George, P.S. and S. Chattopadhyay. 2001. 'Population and Land Use in Kerala, in *Growing Populations, Changing Landscapes. Studies from India, China, and the United States*, ed. Indian National Science Academy, Chinese Academy of Sciences and U.S. National Academy of Sciences. Washington, DC: National Academy Press, pp. 79-105.

Gerritsen, Peter R.W. 2003. 'Back to the Basics: Potentials and Limitations of Socio-Economic Field Methods in 'Fuzzy Empirical Situations'. Input Paper for the 2nd International Training Course of the NCCR North-South. Mexico: University of Guadalajara.

Gill, Gerard J. 1991. *Seasonality and Agriculture in the Developing World: A Problem of the Poor and Powerless*. Cambridge: Cambridge University Press.

Government of India. 2001. Provisional Population Totals (Chapter 7): State of Literacy. Available from: http://www.censusindia.net/data/chapter7.pdf, accessed on 24 January 2006.

Government of Kerala. 2003. *Report of the Commission on WTO Concerns in Agriculture: Building a Sustainable Agricultural Trade Security System for Kerala*. Trivandrum.

Groot, Annemarie. 1998. 'Methods and Techniques for Participatory Problem Solving: A Reference Box. Version 1'. Department of Communication and Innovation Studies, Wageningen Agricultural University, The Netherlands.

Harriss-White, Barbara. 2003. *India Working: Essays in Society and Economy*: Cambridge: Cambridge University Press.

Haupt, Frank, and Ulrike Müller-Böker. 2005. 'Grounded Research and Practice: PAMS—A Transdisciplinary Program Component of the NCCR North–South'. *Mountain Research and Development* 25: 100-3.

Hoi Why Kong. 2002. 'Rubberwood as an Eco-Friendly Source of Tropical Timber'. Paper Presented at Fifth Joint Workshop of the Secretariat of the United Nations Conference on Trade and Development and the International Rubber Study Group on Rubber and the Environment, 5 February 2002, Glasgow, UK.

IFAD. 2001. *Rural Poverty Report 2001: The Challenge of Ending Rural Poverty*. New York: IFAD.

Jacob, James. 2005. 'Forestry and Plantations: Opportunities under Kyoto Protocol'. *Economic and Political Weekly*: 40.

Jalan, Jyotsna and Martin Ravallion. 1999. 'Are the Poor Less Well Insured? Evidence on Vulnerability to Income Risk in Rural China'. *Journal of Development Economics* 58: 61-81.

Joseph, Toms and K. Tharian George. 2002. *WTO and the Natural Rubber Sector. WTO and Natural Rubber Sector in India*, vol. II. Kottayam: Rubber Research Institute of India, Economic Research Division.

Joshi, P.K., Ashok Gulati, Pratap S. Birthal and Laxmi Tewari. 2004. 'Agriculture Diversification in South Asia. Patterns, Determinants and Policy Implications'. *Economic and Political Weekly* 2004: 2457-67.

Kabeer, Naila. 2002. 'Safety Nets and Opportunity Ladders: Addressing Vulnerability and Enhancing Productivity in South Asia'. *Development Policy Review* 20: 589-614.

Kannan, K.P. 2005. 'Kerala's Turnaround in Growth. Role of Social Development, Remittances and Reform'. *Economic and Political Weekly*: 548-54.

Kaspar, Heidi and Michael Kollmair. 2006. 'The Household as an Analytical Category: Concepts and Challenges', in *Gesellschaft und Raum: Konzepte und Kategorien*, ed. Norman Backhaus and Ulrike Müller-Böker. Zurich: Schriftenreihe Humangeographie 22, pp. 103-23.

Kerala State Planning Board. 2002. *Economic Review 2002*. Thiruvananthapuram: Government of Kerala.

——— 2003. *Economic Review 2003*. Thiruvananthapuram: Government of Kerala.

Kershaw, K.L. 1953. 'Mundakayam', in *UPASI 1893-1953*, ed. S.G. Speer. Coonoor: The United Planters' Association of Southern India, pp. 210-29.

KFPE. 2003. *Guidelines for Research in Partnership with Developing Countries. 11 Principles*. Berne: Swiss Commission for Research Partnership with Developing Countries (KFPE).

Killick, Tony. 2001. 'Globalisation and the Rural Poor'. *Development Policy Review* 19: 155-80.

Kochhar, S.L. 1998. *Economic Botany in the Tropics*. New Delhi: Macmillan India Limited.

Koczberski, Gina, and George N. Curry. 2005. 'Making a Living: Land Pressures and Changing Livelihood Strategies among Oil Palm Settlers in Papua New Guinea'. *Agricultural Systems* 85: 324-39.

Krishna, Anirudh. 2005. 'Pathways Out of and Into Poverty in 36 Villages of Andhra Pradesh, India'. *World Development* 34: 271-88.

Krishnakumar, A.K. 1997. 'Rubber', in *The Natural Resources of Kerala*, ed. Thampi Balachandran, N.M Nayar and C.S. Nair. Thiruvananthapuram: World Wide Fund, pp. 228-34.

Kydd, Jonathan and Andrew Dorward. 2001. 'The Washington Consensus on Poor Country Agriculture: Analysis, Prescription and Institutional Gaps'. *Development Policy Review* 19: 467-78.

Lanjouw, Jean O. and Peter Lanjouw. 2001. 'The Rural Non-farm Sector: Issues and Evidence from Developing Countries'. *Agricultural Economics* 26: 1-23.

Lanjouw, Peter, and Abusaleh Shariff. 2004. 'Rural Non-Farm Employment in India. Access, Incomes and Poverty Impact'. *Economic and Political Weekly*. 2004: 4429-46.

Lekshmi, S. and K. Tharian George. 2003. 'Expansion of Natural Rubber Cultivation in Kerala: An Exploratory Analysis'. *Indian Journal of Agricultural Economics* 58: 218-33.

Long, Norman, and Ann Long (eds.). 1992. *The Battlefields of Knowledge: The Interlocking of Theory and Practice in Social Research and Development*. New York: Routledge.

Lutz, Ernst, Hans Binswanger, Peter Hazell, and Alexander McCalla. 1998. *Agriculture and the Environment—Perspectives on Sustainable Rural Development.* Washington D.C.: The World Bank.

Mahesh, R. and V. Vinod. 2005. 'Livelihood Crisis and Responses: Cases from Wayanad'. Paper presented at workshop on 'Livelihood Crises, Coping Strategies and Institutions'. December 2005, Centre for Development Studies, Trivandrum.

Marsh, Robin. 2003. 'Working with Local Institutions to Support Sustainable Livelihoods'. Rome: Rural Development Division, Food and Agriculture Organisation of the United Nations.

Marshall, Catherine. 1994. *Designing Qualitative Research.* Thousand Oaks: Sage Publications.

Mason, Jennifer. 2003. *Qualitative Researching.* London: Sage Publications.

Meert, H., G. van Huylenbroeck, T. Vernimmen, M. Bourgeois and E. van Hecke. 2005. 'Farm Household Survival Strategies and Diversification on Marginal Farms'. *Journal of Rural Studies* 2005: 81-97.

Mehta, Lyla, Melissa Leach, Peter Newell, Ian Scoones, K. Sivaramakrishnan and Sally-Anne Way. 1999. 'Exploring Understandings of Institutions and Uncertainty: New Directions in Natural Resource Management', in *IDS Discussion Paper 372.* Sussex.

Mertz, Ole, Reed L. Wadley and Andreas Egelund Christensen. 2005. 'Local Land Use Strategies in a Globalizing World: Subsistence Farming: Cash Crops and Income Diversification'. *Agricultural Systems* 85: 209-15.

Mikkelsen, Britha Helene. 1995. *Methods for Development Work and Research: A Guide for Practitioners.* New Delhi: Sage Publications.

Mokhtar, Jamil. 2006. *The Fairtrade Potential of the Rubber Supply Chain in Kerala. A Case Study Conducted on Chappals and Condoms.* Masters thesis. Zurich: Department of Geography, University of Zurich.

Müller-Böker, Ulrike. 2001. 'Institutionelle Regelungen im Entwicklungsprozess. Einführung zum Themenheft'. *Geographica Helvetica* 56: 2-3.

——— 2004. 'JACS South-Asia. Sustainable Development in Marginal Regions of South Asia', in *Research for Mitigating Syndromes of Global Change: A Transdisciplinary Appraisal of Selected Regions of the World to Prepare Development-Oriented Research Partnerships,* ed. Hans Hurni, Urs Wiesmann and Roland Schertenleib, Berne: Geographica Bernensia, pp. 225-61.

Mummert, Uwe. 1999. 'Wirtschaftliche Entwicklung und Institutionen. Die Perspektive der Neuen Institutionenökonomik', in *Neue Ansätze zur Entwicklungstheorie,* ed. Reinhold E. Thiel, Bonn, pp. 300-11.

Nair, K.N. and Vineetha Menon. 2004. 'Reforming Agriculture in a Globalising World—The Road Ahead for Kerala'. IP6 Working Paper No. 3. Zurich: Development Study Group.

——— 2006. 'Lease Farming in Kerala: Findings from Micro Level Studies'. *Economic and Political Weekly*: 2732-8.

NCCR North-South. 2002. *PAMS Procedures*. Endorsed by the BoD on 11 September 2002. Berne: NCCR North-South, University of Berne.

Netting, Robert McC. 1993. *Smallholders, Householders: Farm Families and the Ecology of Intensive, Sustainable Agriculture*. Stanford, California: Stanford University Press.

North, Douglass C. 1990. *Institutions, Institutional Change and Economic Performance*. New York: Cambridge University Press.

Nuijten, Monique. 1999. *Institutions and Organising Practices: Conceptual Discussion*. Department of Rural Development Sociology, Agricultural University Wageningen, The Netherlands. Available from: http://www.fao.org/sd/ROdirect/ROan0020.htm, accessed on November 2002.

NZZ. 2002. *Bitterer Kaffee für indische Pflanzer. Historisch tiefe Weltmarktpreise führen zu Selbstmorden*, Neue Zürcher Zeitung— Online, 26 August 2002.

Omamo, Steven Were. 1998. 'Farm-to-Market Transaction Costs and Specialisation in Small-Scale Agriculture: Explorations with a Non-separable Household Model'. *The Journal of Development Studies* 35: 152-63.

Orr, Alastair and Blessings Mwale. 2001. 'Adapting to Adjustment: Smallholder Livelihood Strategies in Southern Malawi'. *World Development* 29: 1325-43.

Patnaik, Utsa. 2002. 'Deflation and *Déjà Vu*. Indian Agriculture in the World Economy', in *Agrarian Studies. Essays on Agrarian Relations in Less-Developed Countries*, ed. V.K. Ramachandran and Madhura Swaminathan. Kolkata: Tulika, pp. 111-50.

Pingali, Prabhu, Yasmeen Khwaja and Madelon Meijer. 2005. 'Commercializing Small Farms: Reducing Transaction Costs'. Paper presented at 'The Future of Small Farms' (research workshop, 26-9 June), Wye, UK. IFPRI, ODI, Imperial College London.

Portmann, Daniel and Balz Strasser. 2005. *Kautschuk als Segen und Herausforderung. Eine bedeutende Einkommensquelle für Keralas Bauern*, Neue Zürcher Zeitung, 14 June 2005.

Prakash, B.A. (ed.). 1994. *Kerala's Economy: Performance, Problems, Prospects*. New Delhi: Sage Publications.

——— 1999. 'Economic Reforms and the Performance of Kerala's Economy', in *Kerala's Economic Development*, ed. B.A. Prakash. New Delhi: Sage Publications, pp. 27-46.

Pronk, Marco. 1997. 'Changing Land Use Practices: Motives and Consequences of Clay Mining and Brick Production in Trichur District, Kerala, India', in *Project Paper II of the Joint CDS-GIUZ Research Project on 'Land Use Dynamics in Kerala: An Actor-oriented Approach*. CDS, Thiruvananthapuram.

Rajasekharan, P. and S. Veeraputhran. 2002. 'Adoption of Intercropping in Rubber Smallholdings in Kerala, India: A Tobit Analysis'. *Agroforestry Systems* 56: 1-11.

Ramakumar, R., K. Narayanan Nair, Vineetha Menon and V.N. Jayachandran. 2005. 'Agrarian Distress and Rural Livelihoods: A Study in Upputhara Panchayat, Idukki District, Kerala State'. Paper presented at workshop on 'Livelihood Crises, Coping Strategies and Institutions', November 2005, Centre for Development Studies, Trivandrum.

Ramisch, Joshua, James Keeley, Ian Scoones and William Wolmer. 2002. 'Crop-Livestock Policy in Africa: What Is To Be Done?' in *Pathways of Change in Africa. Crops, Livestock & Livelihoods in Mali, Ethiopia & Zimbabwe*, ed. Ian Scoones and William Wolmer. Heinemann, pp. 183-225.

Reardon, Thomas. 1997. 'Using Evidence of Household Income Diversification to Inform Study of the Rural Nonfarm Labor Market in Africa'. *World Development* 25: 735-47.

Reardon, Thomas, L. Christopher Delgado and Peter Matlon. 1992. 'Determinants and Effects of Income Diversification amongst Farm Households in Burkina Faso'. *The Journal of Development Studies* 28: 264-96.

Reardon, Tom, K. Stamoulis, M.E. Cruz, A. Balisacan, J. Berdegue and B. Banks. 1998. 'Rural Non-farm Income in Developing Countries, in *FAO: The State of Food and Agriculture*, ed. FAO, Rome: FAO.

Rehm, Sigmund, and Gustav Espig. 1991. *The Cultivated Plants of the Tropics and Subtropics: Cultivation, Economic Value, Utilization.* Göttingen: Verlag Josef Margraf.

Ribot, Jesse C. 1998. 'Theorizing Access: Forest Profits along Senegal's Charcoal Commodity Chain'. *Development and Change* 29: 307-41.

Richards, Paul. 1985. *Indigenous Agricultural Revolution. Ecology and Food Production in West Africa.* London: Hutchison University Library.

Rubber Board. 2003a. *Indian Rubber Statistics*, vol. 26. Kottayam: The Rubber Board, Ministry of Commerce & Industry.

——— 2003b. *Rubber and its Cultivation 2003*. Kottayam: The Rubber Board.

——— 2004. *Rubber and its Cultivation 2004*. Kottayam: The Rubber Board.

——— 2005. *Rubber Statistical News*, February 2005.

Rubin, Herbert and Irene Rubin. 1995. *Qualitative Interviewing: The Art of Hearing Data.* London: Sage Publications.

Sachs, Ignacy. 2000. *Understanding Development: People, Markets and the State in Mixed Economies*. New Delhi: Oxford University Press.

Santhakumar, V. and K. Narayanan Nair. 1999. 'Kerala's Agriculture. Trends and Prospects', in *Kerala's Development Experience*, ed. M.A. Oomen. New Delhi: Concept Publishing Company, pp. 314-24.

SARPI. 2000. *NCCR North-South Research Proposal. Proposal for a National Centre of Competence in Research (NCCR)*. Submitted to the Swiss National Science Foundation on 15 March 2000. Berne.

Scoones, Ian. 1998. 'Sustainable Rural Livelihoods: A Framework for Analysis'. IDS Working Paper 72. Brighton: Institute of Development Studies (IDS).

Scoones, Ian and William Wolmer. 2002. *Pathways of Change in Africa. Crops, Livestock & Livelihoods in Mali, Ethiopia & Zimbabwe*: Heinemann.

Scott, W. Richard. 1995. *Institutions and Organisations*. Thousand Oaks, California: Sage Publications.

Sekhar, C.S.C. 2004. 'Agricultural Price Volatility in International and Indian Markets'. *Economic and Political Weekly* 2004: 4729-36.

Sieberg, Herward. 1999. 'Duale Wirtschaft und ein unbegrenztes Angebot an Arbeitskräften'. *E+Z - Entwicklung und Zusammenarbeit* 1999: 176-8.

Silvermann, David. 2000. *Doing Qualitative Research: A Practical Handbook*. New Delhi: Sage Publications.

Start, Daniel. 2001. 'The Rise and Fall of the Rural Non-Farm Economy: Poverty Impacts and Policy Options'. *Development Policy Review* 19: 491-505.

Strasser, Balz. 2000. 'Financing Rural Infrastructure Projects with Village Participation, Grant and Loan. A Sociological and Financial Analysis of Financing Infrastructure, Operation and Maintenance of a Rural Water Supply Network: The Case Study of Kedjom Ketinguh in the North-West Province of Cameroon'. Masters thesis. Zurich: Department of Agricultural Economics, ETH Zurich.

——— 2004. 'Household Listing: A Short Method Working Paper'. Unpublished. Zurich: Development Study Group.

Stutz, Heidi. 1999. *Distributional Politics and Economic Growth—Insights from Reform Strategies in the Agrarian Sector of Kerala (India)*. Lizentiatsarbeit. Zurich: Abteilung Wirtschaftsgeschichte, University of Zurich.

Syamasundaran Nair, K.N. [nd]. *Agro-Climatic Zones of Kerala*. Updated Excerpts from the Report of the Committee on Agroclimatic Zones and Cropping Patterns: Kerala Agricultural University.

Tellis, Winston. 1997. 'Introduction to case study', *The Qualitative Report (online Journal)*: no pagination.

Thalanadu Grama Panchayat. 2002. *Thalanadu Grama Panchayat Development Report for the 10th Five-Year Plan (2002-2007)*.

The Hindu. 2003. 'Diversification Must in Rubber Sector', *The Hindu*, 21 December 2003.

Thüler, Sue. 2002. 'Seilziehen um wissenschaftliche Repräsentationen des Haushalts. www.ethno.unibe.ch/arbeitsblaetter/AB21_Thu. pdf', in *Arbeitsblatt Nr. 21. Institut für Ethnologie, Universität Bern*. Bern.

Trueb, L. 1996. *Naturkautschuk: Traenen des „weinenden Baums'. Anbau von Hevea brasiliensis in Suedostasien. Ein genuegsamer Baum*, Neue Zürcher Zeitung, 22 May 1996.

U.S. Census Bureau. 2003. 'CSPro Getting Started'. Version 2.4. Downloaded from http://www.measuredhs.com/cspro/docs/start.cfm on Nov. 18th, 2003. Washington DC, p. 39.

Uphoff, Norman. 1993. 'Grassroots Organizations and NGOs in Rural Development: Opportunities with Diminishing States and Expanding Markets'. *World Development* 21: 607-22.

Véron, René. 1998. *Markets, Environment and Development in South India. Cultivation and Marketing of Pineapple and Cashew in Kerala*. Dissertation. Zurich, University of Zurich.

——— 1999. *Real Markets and Environmental Change in Kerala, India. A New Understanding of the Impact of Crop Markets on Sustainable Development.* Aldershot: Ashgate.

——— 2001. 'The 'New' Kerala Model: Lessons for Sustainable Development'. *World Development* 29: 601-17.

Véron, René, Balz Strasser and Urs Geiser. 2004. 'Globalisierung und Agrarproduktmärkte in Kerala. Das Beispiel Cashew und Kautschuk'. *Geographische Rundschau*: 18-24.

Wadley, Reed L. and Ole Mertz. 2005. 'Pepper in a Time of Crisis: Smallholder Buffering Strategies in Sarawak, Malaysia and West Kalimantan, Indonesia'. *Agricultural Systems* 85: 289-305.

White, Howard. 2001. *Combining Quantitative and Qualitative Approaches in Poverty Analysis. Background paper for a presentation to the Global Development Network on May 14th, 2001.* Sussex: Institute of Development Studies, University of Sussex.

——— 2002. 'Combining Quantitative and Qualitative Approaches in Poverty Analysis'. *World Development* 30: 511-22.

Williamson, Oliver E. 1985. *The Economic Institutions of Capitalism. Firm, Markets, Relational Contracting.* New York: The Free Press.

Wilson, Caroline. 2004. 'Understanding the Dynamics of Socio-Economic Mobility: Tales from Two Indian Villages'. Working Paper 236. London: ODI.

Woolcock, Michael and Deepa Narayan. 2000. 'Social Capital: Implications for Development Theory, Research, and Policy'. *The World Bank Research Observer* 15: 225-49.

World Bank. 1975. *Rural Development. Sector Policy Paper.* Washington DC: The World Bank.

——— 2002. 'Institutions for Sustainable Development', in *World Development Report 2003. Sustainable Development in a Dynamic World: Transforming Institutions, Growth, and Quality of life.* Washington DC: The World Bank, pp. 37-58 (Chap. 3).

——— 2002a. *Poverty and Vulnerability in South Asia.* Washington: The World Bank.

——— 2002b. *World Development Report 2003. Sustainable Development in a Dynamic World: Transforming Institutions, Growth, and Quality of Life.* Washington: The World Bank.

——— 2003. *Reaching the Rural Poor—A Renewed Strategy for Rural Development.* Washington: The World Bank.

Xia, Qingjie and Colin Simmons. 2004. 'Diversify and Prosper: Peasant Households Participating in Emerging Markets in Northeast Rural China'. *China Economic Review* 15: 375-97.

Yin, Robert. 2003. *Case Study Research. Design and Methods.* London: Sage Publications.

Zephyr, Frank, and Aldo Musacchio. 2002. *The International Natural Rubber Market, 1870-1930.* Available from: EH.Net Encyclopedia: http://eh.net/encyclopedia/article/frank.international.rubber. market, accessed on 13 January 2006.

Zollinger, Barbara. 2005. 'Female-Headed Rubber Holdings in Kerala. Impact of Fluctuating Natural Rubber Prices on Income Generating and Expenditure Strategies'. Masters thesis. Zurich: Department of Geography, University of Zurich.

Zoomers, Annelies. 2001. 'Land and Sustainable Livelihoods: Issues for Debate', in *Land and Sustainable Livelihood in Latin America*, ed. Annelies Zoomers. Amsterdam: KIT Publishers, pp. 245-57.

# Index